Integrated Genomics

Integrated Genomics

A Discovery-based Laboratory Course

Guy A. Caldwell, Shelli N. Williams and **Kim A. Caldwell**

Department of Biological Sciences
The University of Alabama
Tuscaloosa, Alabama

John Wiley & Sons, Ltd

Other Wiley Editorial Offices

John Wiley & Sons Inc., 111 River Street, Hoboken, NJ 07030, USA

Jossey-Bass, 989 Market Street, San Francisco, CA 94103-1741, USA

Wiley-VCH Verlag GmbH, Boschstr. 12, D-69469 Weinheim, Germany

John Wiley & Sons Australia Ltd, 42 McDougall Street, Milton, Queensland 4064, Australia

John Wiley & Sons (Asia) Pte Ltd, 2 Clementi Loop #02-01, Jin Xing Distripark, Singapore 129809

John Wiley & Sons Canada Ltd, 22 Worcester Road, Etobicoke, Ontario, Canada M9W 1L1

Wiley also publishes its books in a variety of electronic formats. Some content that appears in print may not be available in
electronic books.

Library of Congress Cataloging in Publication Data

Caldwell, Guy A.
    Integrated genomics : a discovery-based laboratory course / Guy A. Caldwell, Shelli N. Williams, and Kim A. Caldwell.
      p.   ;   cm.
    Includes bibliographical references and index.
    ISBN-13: 978-0-470-09501-0 (HB : alk. paper)
    ISBN-10: 0-470-09501-6 (HB : alk. paper)
    ISBN-13: 978-0-470-09502-7 (PB : alk. paper)
    ISBN-10: 0-470-09502-4 (PB : alk. paper)
    1. Genomics—Laboratory manuals. 2. Diagnosis, Laboratory.
    3. Molecular biology—Technique. I. Williams, Shelli N.   II. Caldwell, Kim A.   III. Title.
    [DNLM: 1. Genomics—Laboratory Manuals. 2. Genetic Techniques—Laboratory Manuals. QU 25 C147i 2006]
    QH450.2.C35 2006
    572.8′6078—dc22   2006017997

British Library Cataloguing in Publication Data

A catalogue record for this book is available from the British Library

ISBN-13 978-0-470-09501-6 (HB)   978-0-470-09502-4 (PB)
ISBN-10 0-470-09501-6 (HB)   0-470-09502-4 (PB)

Typeset in 11.5/16pt Times by Integra Software Services Pvt. Ltd, Pondicherry, India

Guy A. Caldwell would like to dedicate this book to his mentor and friend, Jeffrey M. Becker, PhD

Shelli N. Williams would like to dedicate this to her mother, who taught her to be strong

Kim A. Caldwell would like to dedicate this to her beloved bunny rabbits, The Heft and Rustamov

Contents

Preface

Science is about discovery. The *creation* of knowledge from where once there was none. The process of discovery is *the* passion that drives research, is the fuel that has built our society, and spurred great technological advancement and achievement for the human race. Arguably the greatest monument to our kind has been the unraveling of our complete DNA sequence, the human genome. The wealth of underlying secrets and mysteries that scientists may reveal from this scroll of life has only begun to be tapped. The now mature fields of genomics and proteomics are merely conduits for future discovery and expansion of knowledge through the many other realms of cellular and molecular biology and their growing interdependence on computer science, physical science, and biological engineering. Through all this is that underlying thread of life, DNA, and its universal role as the primary code for how cells, microbes, viruses, animals, plants, and humans reproduce and function. Could there be a greater quest or adventure than to investigate the underlying mechanisms of nature?

College and university life is about discovery too. As students progress through their most formative years of self-growth and independence, capturing the essence of the learning experience is often lost in the practical laboratory environment. Although any accomplished researcher can appreciate a finely devised and written experimental protocol, it is clear to even the novice that *actual* discovery is not possible in the context of the 'cookbook recipes' common to laboratory-based classroom experience. We strive to change this widely held paradigm.

Integrated Genomics is an outgrowth of a highly successful laboratory course taught for the past six years at The University of Alabama. This course was originally devised to meet the formative expectations of the Howard Hughes Medical Institute Undergraduate Biological Sciences Education Program Grant awarded to The University of Alabama and has been modified with subsequent support from the National Science Foundation CAREER program. Both of these fine institutions share the common goal of fostering inquiry-based pedagogy and modernizing biological science education. This text provides the instructor with a broad and innovative strategy through which they may present a true *discovery-based* experience for their students. *Integrated Genomics* represents a continuum of flexible yet organized exercises, wherein concepts of basic microbiology, genetics, molecular biology, genomics, proteomics, and bioinformatics are enmeshed within a framework of experiments for which the results are *not* predetermined. The application of simple model organisms such as bacteria, yeast and worms in this course not only keeps the overall cost of implementing the class down, but equally introduces students to the unity of biological function and the power of comparative genomic analysis. Unequivocally, the burden of discovery-based pedagogy is a challenge, but one that is welcomed by the more ambitious of teachers, where the rewards for both the instructor and students are immeasurable. In this regard, the experimental design of *Integrated Genomics* is inherently adaptable and enables the instructor to 'plug and play' with their own favorite genes and proteins. We sincerely hope that this philosophy will be embraced and extended by inventive professors, and we welcome their feedback.

Discovering the process of discovery is what this book is all about. Moreover, it is a process that is addictive – and simply a lot of fun! Enjoy!

You can find further information about *Integrated Genomic* instructional materials and experimental reagents, including all strains and plasmids at www.wiley.com/go/caldwell

Guy A. Caldwell
Shelli N. Williams
Kim A. Caldwell

Author biographies

Guy A. Caldwell, PhD, is an Associate Professor in the Department of Biological Sciences at The University of Alabama, where since 1999 he has held an undergraduate professorial appointment from the Howard Hughes Medical Institute. He holds an adjunct appointment at the University of Alabama at Birmingham, as an Assistant Research Professor of Neurology. In 2001, Dr Caldwell was named a Basil O'Connor Scholar of The March of Dimes Birth Defects Foundation for his research into the molecular basis of childhood birth defects of the brain. Dr Caldwell is a recipient of grants from The March of Dimes, National Institutes of Health, Dystonia Medical Research Foundation, Parkinson's Disease Foundation, National Parkinson Foundation, and the Bachmann–Strauss Dystonia & Parkinson Foundation. In 2003, The Caldwell Laboratory was selected as 1 of only 11 groups worldwide to represent the research goals of The Michael J. Fox Foundation for Parkinson's Research in their Protein Degradation Grant Initiative. For his combined teaching and research efforts, Dr Caldwell was also chosen as the recipient of a 2003 CAREER Award from the National Science Foundation. In 2005, he was named Alabama Professor of the Year by the Carnegie Foundation for the Advancement of Teaching and Council for Advancement and Support of Education. Dr Caldwell, a native of the New York City area, received his undergraduate degree in Biology from Washington & Lee University in 1986 and his PhD in Cell, Molecular & Developmental Biology from The University of Tennessee in 1993. Following receipt of his doctorate, he moved to Columbia University in New York where he was twice named the recipient of fellowships from the National Institute of Neurological Disease and Stroke. He is the author of two editions of a widely adopted textbook, *Biotechnology: A Laboratory Course*, published worldwide in three languages. He teaches courses in Integrated Genomics, Neuronal Signaling, General Biology, and an acclaimed seminar on the societal impact of the Human Genome Project.

Shelli N. Williams, PhD, is a research scientist at a private forensic company based in Virginia. Following her early graduation *magna cum laude* from undergraduate studies, Dr Williams began her graduate work in the laboratory of Drs Guy and Kim Caldwell at The University of Alabama, where she earned her doctorate from The University of Alabama in 2006. Dr Williams served as an adjunct faculty member in New College, an interdisciplinary department at The University of Alabama, where she was the instructor for a seminar course demonstrating how the nature of the laboratory experience plays an essential role in the understanding and advancement of science. She has experience teaching introductory biology courses to both majors and non-majors students and has been a repeated guest lecturer in a cross-disciplinary bioethics class. As a PhD candidate, Dr Williams served as a teaching assistant for *Integrated Genomics*, a discovery-based genomics course funded by the Howard Hughes Medical Institute. Dr Williams was named the recipient of two university-wide Graduate Council Fellowships, as well as receiving recognition as an Isabella Hummel Graham Scholar honoring outstanding female students throughout the university. She also received a competitive Worthington Biochemical Travel Award from the American Society of Cell Biology, placing her among the highest honored student researchers at their 2003 conference. Subsequent graduate work establishing *Caenorhabditis elegans* as a model for epilepsy was highlighted in news releases by the Howard Hughes Medical Institute. In recognition of her accomplishments, Dr Williams was awarded the 2005 Joab Langston Thomas Award, the top honor for PhD students in Biological Sciences at The University of Alabama.

Kim A. Caldwell, PhD, is an Assistant Professor in the Department of Biological Sciences at The University of Alabama. Dr Caldwell is a Faculty Affiliate of The University of Alabama Center for Green Manufacturing and she is an Adjunct Research Assistant Professor in the Department of Neurology at the University of Alabama at Birmingham Medical School. Dr Caldwell, a native of the Buffalo area, received her undergraduate degree in Recombinant Gene Technology from The State University of New York at Fredonia and her MS and PhD degrees in Biotechnology and Cell, Molecular & Developmental Biology, respectively, from The University of Tennessee. While at Tennessee, Dr Caldwell was a four-time recipient of the Oak Ridge National Lab–UT Science Alliance Teaching/Research Award and the Chancellor's Award for Extraordinary Professional Promise. Following receipt of her doctorate, she held postdoctoral research appointments at The Rockefeller University and Columbia University in New York, during which time she was named the recipient of a Revson Fellowship and a National Research Service Award from the National Institute of Child Health and Human Development. Her research has been published in many outstanding peer-reviewed journals, including *Nature, Proceedings of the National Academy of Sciences, Journal of*

Neuroscience, Human Molecular Genetics, the *Journal of Cell Science* and *Development*. Dr Caldwell serves as Director of the Howard Hughes Medical Institute Rural Science Scholars program at Alabama. Additionally, she has designed and taught courses in General Biology, a seminar on the societal impact of the Human Genome Project, and a course entitled 'The Language of Research', which she teaches jointly for Howard Hughes Research Interns at both Stillman College and The University of Alabama. For her teaching efforts, in 2005 Dr Caldwell was selected as a Education Fellow in the Life Sciences of the National Academy of Sciences.

Acknowledgments

The authors would like to acknowledge the significant contributions of several individuals and organizations that enabled this text to move from a concept to a reality.

Significant financial support has come from a Howard Hughes Medical Institute Undergraduate Biological Sciences Education Program Grant awarded to The University of Alabama and a National Science Foundation CAREER Award to Guy A. Caldwell.

Many thanks to Dr Stevan Marcus and Peirong Yang of the M.D. Anderson Cancer Center for generously sharing their advice and yeast two-hybrid reagents. Thanks to Dr Mingxia Huang of the University of Colorado Health Science Center for plasmid vector pACT2.2 and Dr Andy Fire, Stanford University, for additional worm plasmids used in the development of this book. We wish to thank Dr Martin Chalfie of Columbia University and Theresa Steirnagle of the *Caenorhabditis* Genetics Center (University of Minnesota) for providing some of the *Caenorhabditis elegans* strains used herein and Drs Lisa Timmons (University of Kansas) and Lynn Boyd (University of Alabama in Huntsville) for helpful discussions on RNAi feeding methodologies. Thanks to Cody Locke for filming the worm videos associated with this text on-line. We also greatly appreciate the support of Dr Erich Schwarz and his colleagues at Wormbase for advice on bioinformatics and for the continued development of such an outstanding resource.

We are most grateful to all of our past teaching assistants and students, who have generously contributed their thoughtful suggestions toward improvement of this course over the years. Notably, special thanks go to Jafa Armagost for her generous and careful assistance in the final editing of this text. Likewise, special thanks go to all the members of 'The Worm Shack' (Caldwell Laboratory) at The University of Alabama for their cumulative advice, assistance and tolerance (!) through the writing of this text. We wish to sincerely thank our editor at Wiley, Nicky McGirr, for believing in this idea and patiently waiting for it to materialize. Finally, our single greatest thanks go to Dr Martha J. Powell, Chair of The University of Alabama Department of Biological Sciences, who has been our unwavering champion and inspirational leader at every turn in supporting the development and implementation of *Integrated Genomics*.

List of figures

1 Introduction to basic laboratory genetics

Caenorhabditis elegans emerged as a model system in the 1960s, when it was first championed by professor Sydney Brenner and colleagues at the Medical Research Council in London, England. *Caenorhabditis elegans* is an ideal model system for many reasons. It is relatively small (only 1 mm at adulthood), transparent, hermaphroditic, and easy to grow in the lab (it can be treated as a microorganism). Fortunately for researchers, it shares many characteristics with more complex organisms, including, but not limited to, neurons, basic neurotransmitters, hormones, and numerous basic developmental processes.

Caenorhabditis elegans is one of the three lab organisms used throughout this course. 'The worm', as it is fondly known, serves as a whole animal system in which to study the expression of a specific gene (Chapter 2) and the functional consequences of altering its activity (Chapter 8). Prior to embarking on these analyses, it is first necessary to learn basic worm husbandry and handling. *Caenorhabditis elegans* are typically grown in Petri dishes containing media seeded with a bacterial food source (*Escherichia coli*). The media consists of agar and nutrients for both the bacteria and worms.

Before beginning the actual handling of worms, it is important to understand the lifecycle (Figure 1.1) and basic anatomy (Figure 1.2).

Caenorhabditis elegans *development and lifecycle*

Embryo

Following the union of a sperm and oocyte within the gonad of the hermaphrodite, development of the embryo begins. This stage lasts for approximately 12 h, during which

Integrated Genomics   Guy A. Caldwell, Shelli N. Williams, Kim A. Caldwell
© 2006 John Wiley & Sons, Ltd

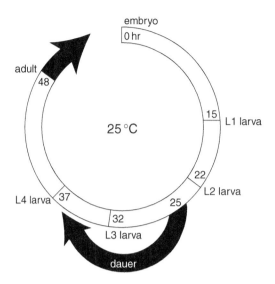

Figure 1.1 *Caenorhabditis elegans* lifecycle at 25 °C

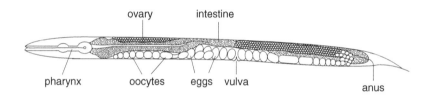

Figure 1.2 Anatomical features of a *Caenorhabditis elegans* hermaphrodite

time the single-celled zygote undergoes round after round of division with no increase in size. What starts as a single cell proceeds to a recognizable, tiny worm within the eggshell (Figure 1.3). At the end of embryogenesis, the larval worm hatches from the egg and crawls away as an L1 stage larva.

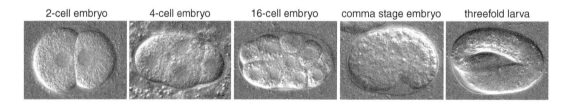

Figure 1.3 Progression through embryogenesis

Larva

During larval development, *C. elegans* increases in size and complexity. The first three larval stages, L1, L2, and L3, are distinguishable morphologically on a dissecting microscope only by differences in size. Examining the worms at higher power magnification will illuminate developmental differences, such as the start of gonadogenesis in the L2 stage and spermatogenesis in the L3 stage. The last larval stage, L4, is easily recognizable in worms because a clear spot in the middle of the ventral side of the animals becomes visible. This clearing is caused by development and formation of the vulval opening (Figure 1.4). *Caenorhabditis elegans* molt between each larval stage, shedding the old cuticle during these events. With some patience *and* a microscope of sufficiently strong magnification, it is possible to observe a worm crawling out of the old layer with hardly a pause in normal activity.

Adult

Adult *C. elegans* are recognizable not only by their larger size, but also by the presence of fully formed gonadal arms and a completely developed vulva. Young adults are distinguishable from older adults by the absence of eggs (Figure 1.4). Adulthood is reached approximately 48 h after initial fertilization of the oocyte by the sperm. Adult worms can live for weeks provided that they have a continuous food source. They typically produce

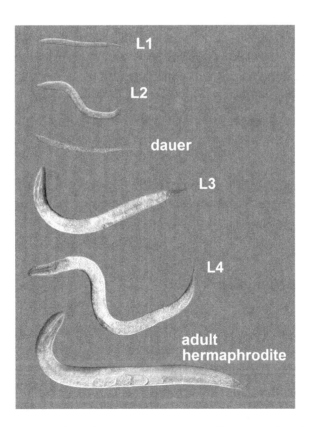

Figure 1.4 Developmental staging of *Caenorhabditis elegans* hermaphrodites from L1 larva to adult

200–300 offspring during their lives. These progeny result from self-fertilization using the sperm and oocytes contained within a single hermaphrodite's gonad.

Alternative larval stage: the dauer larva

Under conditions of stress, such as starvation or severe overcrowding, L2 larvae will not proceed to the L3 stage but rather will arrest development as dauer larvae (Figure 1.4). Dauer larvae are hyperactive, which in the wild would enable them to crawl long distances in search of better conditions. In the dauer stage, *C. elegans* can survive for several months in culture as long as the agar plate they are living on does not desiccate. It is also possible sometimes to recover dauer animals from plates that have desiccated to thin wafers resembling potato chips, although it is not recommended that plates be allowed to dry for this length of time. Dauers will resume normal development when the worms are transferred to a new food source or a plate with fewer animals. The dauer stage is a useful one for two reasons:

- Researchers can maintain stocks often for several weeks without constant supervision or monitoring. Plates are simply allowed to run out of food. Older animals will eventually die but the developing eggs, L1, and L2 larvae on the plate will continue to develop until they reach the dauer stage.

- The dauer stage is a convenient point for synchronizing worm populations. If a plate is allowed to starve, all the worms will either arrest in the dauer stage (younger than L2) or die (L3 and older). Placing worms on a new food source will subsequently cause them to resume development simultaneously, even if they were formed or hatched hours apart.

Can you recognize the different stages? In particular, be sure you can recognize the adult, L4, and dauer stages. These are the most important stages of worms used in this course.

Temperature requirements

Caenorhabditis elegans also display temperature-dependent growth and development. The time line outlined (Figure 1.1) is the canonical lifecycle when worms are grown at 25 °C. Growing worms at higher temperatures is detrimental to their development and survival and should be avoided. However, worms can be grown at cooler temperatures (even at 10 °C), simply resulting in slower development. Worms are routinely grown in incubators set at 15 °C and 20 °C with no adverse effects. Please refer to Appendix III (Figures AIII.2 and AIII.3) for developmental timing at these temperatures.

Worms are handled in a 'semi-sterile' fashion, wherein transfer tools are sterilized and exposure to the outside environment is minimized. In this regard, care is taken to utilize basic aseptic techniques, as would be required for handling microorganisms such as bacteria or yeast.

1.1 Transferring and handling *C. elegans*

Reagents (see Appendix II for recipes)

NGM worm plates seeded with *E.coli* OP50

Wildtype N2 *C. elegans* strain

Equipment

Worm pick (see Appendix III, Figure AIII.1, and directions contained within this appendix)

Bacterial loop

Bunsen burner or ethanol lamp

Dissecting microscope

Worm transfer

The worm pick

Routine manipulation of *C. elegans* in common laboratory practice is performed using a tool referred to as a 'worm pick' (see Appendix III, Figure AIII.1). This is the preferred tool for transferring individuals. A worm pick consists of a small platinum wire attached to a glass Pasteur pipet. The advantage of this instrument is that it can be rapidly flame sterilized. This wire will heat up and cool quickly. Prolonged flaming is not recommended because it melts the glass of the pipet and will kill the animals if the pick is not sufficiently cooled.

Before you begin transferring worms, practice finding the end of the pick under the dissecting microscope. First locate and focus on the tip using low power magnification.

Subsequently become comfortable with finding the pick as you incrementally increase the magnification. You will note that the field of view decreases as magnification increases. Thus it is essential that your pick be in the center of the field of view. Practice transferring worms using adult animals because they are larger and easier to transfer.

Some general points to keep in mind as you practice worm transfer:

- Be very careful not to scratch the agar of the plate. Puncturing the agar will generate a hole that animals are sure to enter and burrow. Once in the agar, they are unrecoverable and therefore cannot be used in an experiment.

- When removing the lid of a worm plate, always place it straight down on the table. Do not flip the lid over so that the surface pointing towards the plate is then pointing towards the environment. This is almost certain to lead to contaminated worm plates.

- See video demonstrating worm transfer. www.wiley.com/go/caldwell

(1) Flame the platinum wire for 3–4 s using a Bunsen burner to sterilize the pick.

(2) Transfer an adult worm using one of two methods:

(a) *The shoveling method.* Think of the pick as a spatula typically used in flipping hamburgers. Position the pick beneath the animal *without puncturing the agar* and lift it off the plate. Transfer the worm to a new plate by placing the edge of the pick onto the top of the agar and sliding the worm off its edge. It is sometimes easier to do this at the edge of the bacterial lawn. This method works fine with practice, but sometimes damages the animals (by accidental stabbing).

(b) *The sweeping method.* Open the fresh plate to which worms are to be transferred and put a little dab of bacteria on the flat end of the recently flamed pick. Place the plate with worms on it on the microscope stage, focus on the animals, and then pick up a worm by touching the top of the animal with the surface of the pick containing the bacteria. *Do not puncture the surface of the agar.* Think of the pick as a paintbrush that gently sweeps across the animal as you attempt to pick it up. The animal will stick to the bacteria. Transfer the worm to the fresh plate by skimming the surface of the agar to help the animal slide off the pick. This method is preferred for beginning students, but one must be careful to transfer *only* the desired animal, as other small larvae or eggs often get stuck to the bacteria on the pick and must be removed afterwards (by the same method).

In both cases, transfer of animals must be performed relatively rapidly (~20s from plate to plate) because the animals will desiccate in the open air. Likewise, if using the sweeping method, prolonged exposure to the air increases the chances of contamination. Additionally, the bacteria will begin to dry and harden and worms will no longer adhere to it. It is essential to become proficient in transferring animals, which will take some amount of practice. It is a good idea to spend several hours on two or more occasions transferring worms before you need to actually transfer them for an experiment. Additionally, this will give you a chance to evaluate your sterile technique.

The bacterial loop

Larger numbers of animals, especially dauer worms, can be transferred using a flame-sterilized bacterial loop. The loop is heated until it glows under a flame and then cooled by plunging into the agar of the worm plate from which you will be transferring worms. The loop is used to strip a small portion of the agar from the plate along with the worms contained on it. The strip is placed gently on top of the agar surface of a fresh plate, *worm side down*, enabling the worms to crawl off it toward the new bacterial lawn.

Try the following exercises:

(1) Transfer three worms each of the L2, L4, and adult stages to a fresh worm plate seeded with bacteria. Have your instructor verify the stages transferred and that you have not punctured the agar.

(2) Try picking up and transferring multiple worms with your pick at the same time.

(3) Isolate a single worm from within a close group of other worms. Can you grab only your target worm, or did you get some 'hitchhikers'?

After you are comfortable with stage identification and handling, practice transferring worms with increasing speed. An average beginning speed is about one worm per minute. An experienced worm researcher can transfer individual animals at a rate of one animal per 10–20 s. Prior to completion of this exercise, it should be your goal to place four L4 stage animals on a fresh plate within 5 min without stabbing the agar. Allow these worms to grow until the next laboratory meeting to verify your sterile technique as well as the viability of animals after you transferred them. Worm plates are incubated agar side up to prevent condensation from forming on the lid and subsequent desiccation of the plate.

Bioinformatics exercise

Spend some time familiarizing yourself with key *C. elegans* websites:

www.wormbase.org   Wormbase is the repository for nearly all available genomic and proteomic information and research being conducted on *C. elegans*. Try a literature search on your favorite gene to find work that others have done on it.

http://elegans.swmed.edu/   The *C. elegans* WWW server is the home of all kinds of helpful information, including protocols, a link to the *C. elegans* Genetics Center (CGC), and bibliographic information on worm research. There are also several links to sites introducing beginners to *C. elegans*, and lab websites of most worm researchers around the world.

www.wormatlas.org   Wormatlas contains anatomical information about the worm, and a searchable database for information on individual cells and tissue types.

www.wormbook.org   Wormbook is the latest version of *C. elegans* information available to the research community. It includes discussions of the biology of the worm as well as protocols and technological advances.

1.2 Introduction to laboratory genetics

As mentioned in the previous section, *C. elegans* are primarily found in the hermaphroditic form, wherein self-fertilization yields offspring. Self-fertilization has several advantages in genetic analysis. Because the act of mating is not required, mutations that might physically disrupt this process can still be maintained (for example, some animals are severely uncoordinated or even paralyzed and therefore incapable of mating, yet can still produce progeny). Stocks of animals are also easily maintained over time because one animal can give rise to an entire population without the necessity for any crosses. Furthermore, if it is necessary to check the progeny of a subsequent generation of animals for a specific phenotype, this is easily done by plating individual worms onto separate agar plates and then allowing them to lay eggs. For example, when trying to obtain or discern homozygotes $(-/-)$ from heterozygotes $(+/-)$, no additional 'crosses' are necessary. Plating individual heterozygotes will automatically yield some homozygous animals in the next generation following self-fertilization.

However, not all worms are hermaphrodites. In fact, a small percentage (~ 0.05 percent) of animals within a worm population will be male. Genetically, both males and hermaphrodites have five pairs of autosomes (I, II, III, IV, and V); however, although hermaphrodites have a pair of X-chromosomes, males only have one copy. A random non-disjunction event during meiosis in the egg or sperm causes the resulting embryo to be an XO male rather than an XX hermaphrodite.

Males can fertilize hermaphrodites, yielding approximately 50 percent hermaphrodite and 50 percent male offspring. Male offspring are therefore the result of cross-fertilization. Hermaphrodites may result from either cross- or self-fertilization. This ability to cross-fertilize using males enables researchers to study the genetic inheritance of traits such a dominance or sex-linkage patterns. It is important to be aware, however, that in all crosses with *C. elegans* one must be able phenotypically to account for self-progeny versus cross-progeny because this dramatically affects genetic analyses. Often, recessive mutation markers are employed to identify self-progeny.

Males can be distinguished easily from hermaphrodites by their tail morphology (see Figure 1.5). Whereas the hermaphrodites have a sleek or tapered tail, male tails end in a hook-like or arrow-shaped structure. This difference first becomes evident after about the L3 stage of larval development and is more pronounced as the males mature to adults. In adult males the 'hook' is quite obvious. Young males may appear to have a visible but less pronounced rounding at the tip of their tails.

Caenorhabditis elegans researchers have had over 40 years to develop mutants from which all the community can draw upon to facilitate genetic analysis. The mutations used as phenotypic markers are often easily recognizable traits. Morphology markers are often used, such as:

- Dumpy (Dpy) phenotype – animals are shorter and fatter than wildtype.

- Multivulva (Muv) phenotype – animals have extra pseudo-vulvae, visible as protrusions along the ventral side of the body.

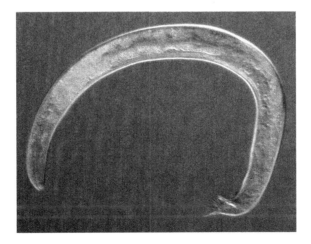

Figure 1.5 *Caenorhabditis elegans* adult male

Markers that cause defects in worm movement or response are also used:

- Mechanosensation (Mec) phenotype – animals do not respond to a light touch stimulus.

- Uncoordinated (Unc) phenotype – animals show defects in movement ranging from small changes in the normal sinusoidal movement patterns to the complete inability to move.

Researchers may also use more complex phenotypic markers, such as the variable abnormal (Vab) phenotype, which affects multiple organ systems within the worms. A more comprehensive (although not exhaustive) listing of worm phenotypes can be found in Appendix IV.

Procedural overview

After you have become familiar with the Mec phenotype, you will attempt to determine the inheritance pattern of a particular *mec* mutation. This will be done by crossing wildtype males into worms of unknown genotypes. In a few days, you will examine the offspring of these crosses and hopefully determine how your mutations are inherited.

Reagents (see Appendix II for recipes)

NGM worm plates seeded with normal-sized *E.coli*: OP50 lawns

NGM worm plates seeded with smaller *E.coli* OP50 mating lawns

Equipment

Eyebrow hair glued to the end of a toothpick

100 percent ethanol in a sterile 1.5-ml microcentrifuge tube

Worm pick

Bunsen burner or ethanol lamp

Dissecting microscope

Assaying touch response

In this exercise, you will first examine a simple characteristic, the 'touch' or Mec (mechanosensory) phenotype. The Mec phenotype is caused by mutations within mechanosensation genes, resulting in worms that will not respond properly to a light touch on the head or tail. The touch is administered by lightly brushing the head of the animal with an eyebrow hair that has been glued to one end of a toothpick. The wildtype response to such a tap is a rapid backing up. If you then touch the worm on the tail, it will then quickly resume its forward motion. Repeating the head touch should result in another reversal. Mec mutants, as they are often called, will not respond with this back–forward–back motion when touched in this sequence.

(1) Obtain several plates of *mec* mutants from your instructor.

(2) Carefully focus on a single animal at a medium range magnification (i.e. 25×).

(3) Dip the eyebrow hair carefully in the 100 percent ethanol to sterilize it. Allow the hair to dry for 5–10 s before touching a worm. *Do not flame sterilize the eyebrow hair*!

(4) Perform the touch assay: very carefully brush your eyebrow hair across the head of the animal, then the tail, and finally the head again.

(5) During each touch, observe the animal carefully to see if it responds to the stimulus by moving in the opposite direction *each* time you touch it.

(6) Repeat the assay with 20 animals to be sure that you can recognize a Mec phenotype. For comparison, perform the touch assay on wildtype animals and notice the difference in response. If bacteria accumulate on the end of the eyebrow hair, occasionally re-sterilize in ethanol, as above.

Genetic crosses to determine the inheritance pattern of unknown mutations

Mating of *C. elegans* males and hermaphrodites is typically performed by placing a few of each onto agar plates seeded with small bacterial lawns and incubating the animals for 2–3 days at 20–25 °C until progeny hatch. Standard procedure is to pick 3–4 hermaphrodites onto a fresh plate first and then add 7–9 young adult males to the same plate. This ratio of males to hermaphrodites is commonly used because of the tendency of male worms to wander off the plates unless encountering a hermaphrodite. It is best to use hermaphrodites that are in the L4 stage of larval development (easily distinguished by a clear patch in the center of the ventral side of the animal). Animals of this stage retain their full sexual potential and maximize the number of cross-progeny generated because the hermaphrodites should not yet have begun self-fertilizing. It is best to transfer animals to the outside edge of the bacterial lawn to visibly ensure that no undesired eggs or additional animals have been transferred accidentally. It is essential to avoid transfer of extraneous embryos or animals because they will produce self-progeny that will 'contaminate' the genetic cross-progeny.

Mating is an inefficient process in *C. elegans* (in comparison to self-fertilization) and, generally, several sets (two or three) of mating plates are performed for *each* desired cross. If using a specific phenotypic mutation that might affect mating (Unc, Dpy), more mating plates should be prepared to achieve successful crosses. Adult males can be removed after 24–48 h to avoid confusion with cross-progeny. The progeny of a male and hermaphrodite cross will yield ~50 percent males and 50 percent hermaphrodites among the approximately ~200 eggs laid per parent. Often one or more hermaphrodites from a plate of several mating worms will not be cross-fertilized and this ratio is skewed toward a higher number of hermaphrodites, generated by self-fertilization.

Test crosses

Day 1

(1) You will be given sets of 'unknown' plates labeled A, B, and C. These plates will contain a predominantly hermaphrodite population – some sets will be *mec* mutant animals. You will also be given a few plates containing a high percentage of wildtype males from wildtype mating crosses.

(2) Set up three mating plates for each set of hermaphrodites (A, B, and C) with wildtype males. This is nine mating plates in total. As noted above, use a ratio of 3–4 hermaphrodites with 7–9 males for each mating plate. Care should be taken not to break or scratch the agar surface because animals will burrow and prevent mating.

(3) Incubate plates inverted at 20 °C for at least 24 h. After this time, your instructor may elect to shift the growth of these animals to a lower or higher temperature. Ideally, you wish to score progeny when they are young adults because it is difficult to score accurately the Mec phenotype by touching younger animals.

3–5 Days later

(4) Carefully examine each of your mating plates for evidence of a moderate percentage (15–30%) male worms in the progeny. Score plates as either "+/+", "+/−" or "−/−" for representing approximate mating success based on the prevalence of males.

(5) Score the progeny (∼20 animals) of each plate that exhibited successfully (+/+ or +/−) plates mated animals for wildtype versus mutant phenotypes. Record these data in your notebook and analyze as requested by the instructor.

   (a) Which sets contain mutant animals in the progeny?

   (b) Do any of these mutations appear to be *dominant*?
       If so, which set(s)? Why?

   (c) Are any of the mutants *sex-linked* or not? How do you know?
       If so, is it recessive sex-linked? How can you tell? If the trait is recessive, what experiment could you perform to tell if wildtype animals are heterozygous or homozygous?

Bioinformatics exercise

Once your instructor tells you what precise Mec mutant you were working with, use Wormbase or the WWW server to determine if your results match what is known about the inheritance of your *mec* gene.

Instructor's notes

Although this text outlines the use of the Mec phenotype to introduce students to *C. elegans* genetics, you are free to choose your own phenotype to study. If you decide to use a different phenotype, keep the following points in mind.

- Is the phenotype easy to identify?
- Are there mutant strains available that show different inheritance patterns?
- Will the mutation itself have an impact on the ability of the worms to mate?

Strains may be acquired from the *C. elegans* Genetics Consortium (CGC) at http://biosci.umn.edu/CGC/CGChomepage.htm. If you order a strain from the CGC, you will be provided with what information is known about the genetic inheritance of the mutation and if any problems are to be expected when using this strain in mating crosses. More information is generally available if one contacts the laboratory that generated the mutant strain of interest. To select strains, simply perform a search for the phenotype (i.e. Mec) and you should see all strains showing that phenotype. You may also acquire wildtype males from the CGC. It will be necessary for you to expand the male population before the test crosses by mating wildtype males to wildtype hermaphrodites. Typically 3–5 plates of males for each student are sufficient to set up all the test crosses. Set up mating plates between wildtype males and hermaphrodites at least 1 week before test crosses are to be set up. The day before the lab, strip dauer animals from the unknown strains and the wildtype cross plates onto 2–3 fresh plates per strain (3–5 for the males) and incubate at 25 °C to ensure that the animals are at the proper stage for setting up the mating plates. It is advisable to monitor developmental stage and shift temperatures appropriately to maximize obtaining L4 stage animals at the time of class. Review the protocols in Appendix III for information on maintaining worm stocks.

During the class, be sure to tell students not to touch the nose of the animals while performing touch assays. The anterior of the animal contains a different set of touch ('head touch') receptors and triggering this response will result in false-positive data. It is common for *mec* animals to respond to the initial touch on the nose but they will not respond to the repetitive head–tail–head touch.

Following the lab, monitor the progress of the mating plates and perform appropriate temperature shifts (colder to slow down) to ensure that the progeny of the crosses do not run out of food and are ideally near the young adult stage when students assay for the Mec phenotype. Discuss the impact of self-progeny in the resultant analysis of the test cross-progeny. It may be helpful for students to draw hypothetical Punnet squares for the various scenarios of a dominant or recessive mutation.

2 Gene expression analysis using transgenic animals

Although *in vitro* assays can give researchers numerous pieces of information regarding a gene or protein's function, no concrete conclusions about its actual role in the cell can be drawn without *in vivo* experiments. A key aspect of studying the function of a specific gene is to determine its expression pattern. Several methods allowing scientists to determine the expression of their favorite gene are available. This chapter details whole animal expression analysis of both fixed (*lacZ*) and live (GFP) animals (Figure 2.1). An overview and the advantages of the two main methods are highlighted at the start of each section below.

Conducting an expression analysis on a given gene initially requires identification of a putative promoter region and generation of an expression construct. The promoter of the gene of interest must be cloned into an expression vector containing the reporter gene (a map of the vectors commonly used by *C. elegans* researchers can be found at the end of this chapter). Tissues that normally express the gene of interest will contain the proper regulatory sequences and proteins that will bind to the gene promoter. Because the reporter gene is now being controlled by the target gene promoter, it will be turned on at the same time and location as that gene.

Before an expression pattern can be analyzed, the construct must be introduced into the organism under study. In the case of *C. elegans*, this is frequently performed by microinjection or, increasingly, by microparticle bombardment. The expression construct must be introduced with a phenotypic marker to facilitate selection of transgenic animals.

Integrated Genomics   Guy A. Caldwell, Shelli N. Williams, Kim A. Caldwell

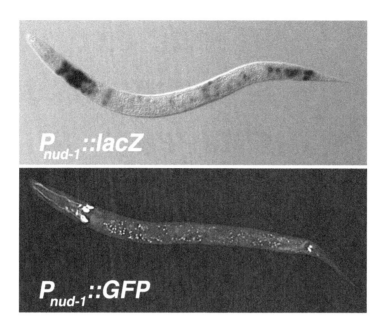

Figure 2.1 Comparison of reporter gene expression results using either *lac*Z or GFP driven by a specific promoter fusion

Caenorhabditis elegans researchers often use a dominant marker (*rol-6*, which encodes a defective collagen gene) that causes the worms to move in a 'rolling' motion rather than their normal sinusoidal locomotive pattern. Worms expressing the roller marker must be maintained at each generation because there is no selective pressure applied to maintain the transgenes within the worm.

When the expression construct and phenotypic marker plasmids are simultaneously introduced into a worm by microinjection, these circular DNAs are nicked and linearized by endogenous nucleases within the worm. Homologous regions within the vectors recombine to form tandem arrays, or long stretches of DNA containing repeated copies of each sequence. The genes comprising these arrays are randomly placed in concatamers and this arrangement often results in automatic overexpression. Although this is standard practice in the worm field, multiple copies of a given gene can be detrimental because overexpression can cause a variety of mutant phenotypes, including lethality.

There are basically two ways by which transgenes can be maintained in worm stocks: as extrachromosomal arrays (subsequently known as stable lines) and as chromosomal insertions (known as integrated lines). Stable lines carry the tandem array as an extra-chromosomal sequence that will be transmitted to some but not all of the offspring of an animal (10–90 percent and typically averaging 50 percent inheritance: Mello and Fire, 1995). Stable lines must be propagated carefully to maintain animals expressing the phenotypic marker. Otherwise, a line may be lost over time because inheritance is not occurring at 100 percent. Integrated lines on the other hand contain arrays that have been inserted directly into the genome. Integration of an array may occur spontaneously (very rare) or can be induced experimentally by causing random DNA breaks that will be sealed by cellular repair machinery. When the breaks are repaired, occasionally the tandem array is incorporated within the chromosome. It will then be inherited by all offspring because it is maintained as a part of the genome.

This exercise is intended to demonstrate where, and possibly when, the specific target gene chosen for analysis is expressed. Knowing which cell types express the target gene can provide an initial framework for hypothesizing what types of interacting proteins may be identified in the subsequent protein–protein interaction screen.

This lab also provides valuable experience with studying *in vivo* expression patterns. Preliminary knowledge of the cells expressing the target gene can provide clues to the functional consequences of knocking down a specific transcript by RNA-mediated interference (RNAi, Chapter 8). For example, identification of a target gene with vulval muscle expression may point to a role in the egg-laying machinery. Subsequent phenotypic analysis may support the observed expression, in this case resulting in an Egl (egg-laying defective) phenotype.

2.1 Transgenic gene expression analysis in *C. elegans*: *lacZ* staining

One method of gene expression analysis makes use of the bacterial *lacZ* gene, encoding the enzyme β-galactosidase (β-gal). β-galactosidase normally acts on the substrate lactose, although it can also cleave 5-bromo-4-chloro-3-indolyl-bD-galactoside (X-gal), a colorless synthetic analog of lactose. When β-gal cleaves X-gal, it generates water-insoluble dichloro-dibromo indigo, which is visible as a bright blue color within the cells where cleavage has occurred.

The main advantage of a *lacZ* assay over *in vivo* assays using fluorescent reporters is that no special illumination sources are needed to visualize the blue precipitate. When the target gene is expressed, the *lacZ* gene also will be expressed, producing β-gal. Following exposure to a solution containing X-gal, simple transmitted light shone through the transparent worm via bright-field microscopy will show where the β-gal-catalyzed reaction has taken place. A second advantage of the *lacZ* system in *C. elegans* is the presence of a nuclear localization signal in the vectors typically used for such studies. This signal targets the reporter directly to nuclei of cells expressing β-gal; pinpoint expression in a nucleus makes it much easier to identify the precise cell showing expression. However, keep in mind that the blue precipitate does diffuse into the cytoplasm at times, resulting in a non-nuclear staining.

If one's lab has access to a microscopy imaging system, it is quite easy to acquire numerous images of the expression pattern of the target gene. Your instructor has already acquired a *lacZ*-containing strain; this strain also co-expresses the *rol-6* roller marker gene, allowing you to isolate transgenic worms by their movement pattern.

Procedural overview

You will begin the lab by carefully washing your worms to remove the bacteria they have been living and feeding upon. This is important not only for the general aesthetics of imaging but also because most bacteria produce β-gal. You will then gently desiccate the worms and crack their cuticles by incubating them in acetone, followed by air-drying for a few minutes. This step is followed by equilibrating the worms to a pH appropriate for β-gal activity by washing them in a mild phosphate buffer (PMB). Worms are then placed in a β-gal staining solution. Most of the components of the staining solution serve as a buffer for the worms and β-gal enzyme. Cyanide is used to stabilize the enzyme and X-gal is, of course, the substrate upon which the enzyme acts. The experimental procedure is designed to maintain intact worm morphology while enabling penetration of the staining solution. Some variability in staining due to differences in drying and the efficiency of cracking the cuticle will occur, causing differences in the quality of the expression pattern observed.

Reagents (see Appendix II for recipes)

Double-distilled water, sterilized by autoclaving

Acetone

PMB (mild phosphate buffer)

Worm β-Galactosidase staining solution (*Be careful – contains cyanide!*); covered in foil (light-sensitive)

Agarose (powder)

1× TAE buffer

Equipment

Pasteur pipet (glass)

Centrifuge tubes, 15 ml (i.e. Falcon 2097)

Tabletop centrifuge

Microcentrifuge tubes, 1.5 ml (sterile)

Platform rocker (i.e. Nutator)

Aluminum foil

Glass slides

Glass coverslips, 22 mm × 22 mm

60 °C Waterbath

Glass flask with boiling cap

Microwave

Differential interference contract (DIC) microscope with 40 × objective (100 × optional)

Camera (digital or 35 mm mounted on microscope)

Before start of the lab

(1) Grow up 3–4 plates of worms containing a transgenic extrachromosomal array encoding a specific *lacZ* gene fusion co-expressed with a dominant marker gene *rol*-6. Worms should be incubated (20–25 °C) until progeny appear and densely cover the plates (preferably just before food has been cleared; do *not* allow worms to grow to point of starvation and dauer formation). Place 5–6 L4 worms on each plate and grow for 5 days if grown at 20 °C.

Day 1

(2) Wash the worms off the plates with sterile water by pipeting 2 ml onto one of the plates, then mixing up and down with a pipettor or glass Pasteur pipet and transferring them with this pipette to a 15-ml centrifuge tube. Repeat for each of the plates – combine the worms into the one tube. Glass Pasteur pipets are preferred because worms will adhere to plastic.

(3) Add water to 8-ml final volume. Pellet worms in a clinical centrifuge for 2 min at 1500 rpm. Carefully remove most of the supernatant with a pipettor.

(4) Wash the worms with 5 ml of water and centrifuge again, as above.

(5) Remove most of the water using a pipettor, leaving approximately $250-300\,\mu l$ of worms and water. Carefully resuspend and transfer these worms to a 1.5-ml microfuge tube using a glass Pasteur pipet.

(6) *Ensure that gloves are worn for the remainder of the experiment.* Add $500\,\mu l$ of acetone to the worms and place the tubes on a rocker for 10 min at room temperature.

(7) Allow the worms to settle for about 5 min. Using a pipettor, carefully remove *as much of the liquid as possible* without taking up worms. Dry the worms by opening the tube lid and leaving upright in a tube rack on the benchtop for 20–30 min.

(8) Add $500\,\mu l$ of PMB to the tube and gently resuspend the worms by flicking the tube with a fingertip to wash them. Allow the worms to settle (~ 10 min), then remove the PMB using a pipettor.

(9) Suspend the worms in $200\,\mu l$ of worm β-gal staining solution and wrap the tubes in foil. Incubate at room temperature for 24 h on the platform rocker. *Be very careful with the staining mix because it contains cyanide*!

(10) Place worms at $4\,°C$ until ready to observe.

Day 2

(11) Allow worms to settle to the bottom of the tube and then remove the staining mix and *discard in hazardous waste container*. Resuspend the worms in $300\,\mu l$ of sterile double-distilled water.

(12) Make a 2 percent agarose solution by melting 1 g of agarose in 50 ml of $1\times$ TAE buffer in the microwave. When the agarose is completely in solution, place it in a $60\,°C$ waterbath.

(13) Mount your worms for microscopic analysis. To prevent worms from being crushed when you place them on a slide, use an agarose pad for support. Place a strip of laboratory labeling tape lengthwise onto two separate glass slides. Place these two slides, tape-side up, on either side of a third slide on a flat surface. Drop a small amount of agarose onto the blank slide in the middle. Quickly place another clean slide perpendicularly across the top of the three slides. The tape on the outer slides determines how thick your agarose support pad will be. After 30–45 s, you should be able to lift the top slide off, leaving an agarose pad on the bottom slide.

Figure 2.2 Example of the method for making an agarose pad for use in mounting worms for microscopy

Be careful – an agarose pad that is too thick will obscure visualization of the worms. (see Figure 2.2 for clarification.)

(14) Flick the tube with your fingers to resuspend the worms and then pipet a small amount of liquid (20 μl) from the tubes. Place this drop containing stained worms onto the microscope slide with the ágarose pad. Carefully lower a coverslip onto the sample to avoid air bubbles. The worms are now ready for microscopic examination.

(15) If your lab has access to an imaging system, obtain images of your stained worms. Choose representative photos to place in your laboratory notebook. Ideally, animals of different sizes (i.e. developmental stages) should be documented. If you do not have access to a digital imaging system, traditional photography using 35-mm film or your own artistic talents may be exploited to document your expression pattern.

Bioinformatics exercise

Although experience with worm anatomy is helpful, it is not essential to be a worm expert in order to ascertain the general expression pattern of a target gene. A wonderful resource for worm anatomy, called Wormatlas, is available to help with identifying cells or organs. Visit Wormatlas at www.wormatlas.org or Wormbase at www.wormbase.org to identify the cells in which you are seeing the blue precipitate from your *lacZ* staining. In many cases, the precise cells expressing the construct will be difficult to identify. However, cell or tissue types (i.e. neuron versus intestine) are readily distinguishable using these databases. Examining multiple expression patterns for multiple genes will aid greatly in learning to identify cells.

Once you can identify the cell types expressing the target gene, ask yourself the following questions:

(1) Does your expression pattern at least partially match what is known about your gene and its function?

(2) Does it appear to change in different stages of development?

2.2 Transgenic gene expression analysis in *C. elegans:* GFP analysis

Green fluorescent protein (GFP) is a naturally fluorescent protein from the jellyfish *Aequorea victoria*. The protein absorbs the energy of UV light and, through chemical changes within the protein, releases photons in the visible (green) wavelength. The most important advantage of GFP is that it can be used in live animals to observe expression patterns for genes. Unlike *lacZ* analysis, GFP can be used to illuminate long cellular processes, such as neurons, and changes in expression can be monitored throughout the life of a single animal. Use of GFP as a reporter gene was first pioneered by Martin Chalfie and colleagues at Columbia University in 1994 using *C. elegans*. Worms are ideal for GFP transgene analysis due to their transparent anatomy.

Another significant advantage of GFP-based assays is the availability of color variants for the protein. The specificity of the emitted light (in the case of GFP wavelengths ranging from 490 to 515 nm) has, in turn, enabled researchers to create different color variants of GFP that fluoresce at different wavelengths. Researchers also have subsequently made small amino acid modifications to increase the signal and stability of the proteins, leading to 'enhanced' fluorescent molecules. Some common examples are blue (eBFP, 415–435 nm), cyan (eCYP, 440–480 nm), and yellow (eYFP, 500–530 nm). A red (DsRed, 575–590 nm) fluorescent protein from coral that has been shown to exhibit excellent stability and reduced quenching is also being utilized by scientists. Researchers are continually developing new variants, allowing for diversity in live visualization. By using fluorescent proteins with different absorbance properties a researcher can tag multiple proteins in a single organism or cell. Exposing the animal to a specific wavelength will cause the fluorescence of only a single protein of a given 'color' at a time.

Additionally, unlike *lacZ* lines, there may be no need for the use of an additional phenotypic marker to select for transgenic animals. Access to a fluorescent dissecting microscope allows researchers to use the actual GFP expression as a marker for the identification of transgenic animals. However, co-injection of a phenotypic marker such as *rol-6* allows researchers to select transgenic worms without a fluorescent dissecting microscope. The largest disadvantage associated with GFP-based analyses (as opposed to *lacZ*) is the need for an expensive fluorescent microscope for proper illumination of the transgenic animals. Fortunately, increasingly inexpensive models are now more widely available and most labs without their own fluorescent microscope will be near a lab that does own one.

Your instructor has either acquired or made a specific GFP-containing strain of *C. elegans*. He/she will tell you whether your transgenic animals carry an additional phenotypic marker and, if so, which one. Additionally, as you work to identify the cells expressing the GFP fusion protein, keep in mind that, unlike the *lacZ* construct in worms, the most commonly used GFP reporter may or may not contain a nuclear localization signal, therefore you will often see diffuse staining in an entire cell rather than localized expression in the nucleus.

Procedural overview

The key reason for using GFP for gene expression analysis is the ability to examine living animals without lengthy preparation. The absence of harsh chemical treatment and fixation reduces variability in the clarity of an expression pattern. One caveat to live animal analysis is that worms move if they are still alive, and thus are more difficult to photograph, therefore a paralysis agent called levamisole is often used to prevent movement during analysis. Levamisole is an acetylcholine agonist that causes contraction of worm body-wall muscles, resulting in overall paralysis.

Reagents (see Appendix II for recipes)

3 mM Levamisole (*Be careful; wear gloves; paralysis agent*)

Agarose (powder)

$1 \times$ TAE buffer

Equipment

Worm pick

Glass slide

Glass coverslip, 22 mm $\times$ 22 mm

60 °C Waterbath

Microscope with appropriate filters for epifluorescent analysis and equipped with a 40 × objective (100 × optional)

Camera (digital or 35 mm mounted on microscope)

Microwave oven (or heat plate)

Glass flask and boiling cap

Before start of the lab

(1) Grow up 3–4 plates of worms containing a transgenic extrachromosomal array encoding a specific GFP gene fusion. Your instructor will inform you of how the extrachromosomal array is phenotypically marked in this strain; be sure you transfer the correct worms. Worms should be incubated (20–25 °C) until progeny appear and densely cover the plates (preferably just before food has been cleared; do *not* allow worms to grow to the point of starvation and dauer formation). Place 5–6 L4 worms on each plate and grow for 5 days if grown at 20 °C.

Day of the lab

(2) Place a 10-μl drop of 3 mM levamisole onto a glass coverslip.

(3) Carefully pick 10–20 worms from the plates you set up a few days ago and place them in the drop of levamisole. You need to do this quickly, within 30–60 s, so that the drop does not dry up.

(4) Make a 2 percent agarose solution by melting 1 g of agarose in 50 ml of 1× TAE buffer in the microwave. When the agarose is completely in solution, place it in a 60 °C waterbath.

(5) Mount your worms for microscopic analysis. To prevent worms from being crushed when you place them on a slide, use an agarose pad for support. As depicted in Figure 2.2, place a strip of laboratory labeling tape lengthwise onto two separate glass slides. Place these two slides, tape-side up, on either side of a third slide on a flat surface. Drop a small amount of agarose onto the blank slide in the middle. Quickly place another clean slide perpendicularly across the top of the three slides. The tape on the outer slides determines how thick your agarose support pad will be. After 30–45 s you should be able to lift the top slide off, leaving an agarose

pad on the bottom slide. Be careful – an agarose pad that is too thick will obscure visualization of the worms.

(6) Invert your coverslip with worms (so that the worms are facing down toward the agarose pad) and carefully drop it onto the agarose pad.

(7) If your lab has access to an imaging system, obtain images of your stained worms. Choose representative photos to place in your laboratory notebook. Ideally, animals of different sizes (i.e. developmental stages) should be documented. If you do not have access to a digital imaging system, traditional photography using 35-mm film or your own artistic talents may be exploited to document your expression pattern.

Bioinformatics exercise

Although experience with worm anatomy is helpful, it is not essential to be a worm expert in order to ascertain the general expression pattern of a target gene. A wonderful resource for worm anatomy, known as Wormatlas, is available to help with identifying cells. Visit Wormatlas at www.wormatlas.org or Wormbase at www.wormbase.org to identify the cells in which you are seeing the fluorescence. In many cases, the precise cells expressing the construct will be difficult to identify. However, cell or tissue types (i.e. neuron versus intestine) are readily distinguishable using these databases. Examining multiple expression patterns for multiple genes will aid greatly in learning to identify cells.

Once you can identify the cell types expressing the target gene, ask yourself the following questions:

(1) Does your expression pattern match what is known about your gene and its function?

(2) Does it appear to change in different stages of development?

Instructor's notes

It is the instructor's prerogative as to whether students perform *lacZ* or GFP analysis, or even both. The availability of a specific transgenic strain may, by necessity, dictate this choice. Transgenic worm strains that may be used for analysis in your course are available free of charge from the *C. elegans* Genetics Consortium (CGC) at http://biosci.umn.edu/CGC/CGChomepage.htm. You can search this site for both *lacZ* and GFP strains. You may initially want to visit Wormbase to choose a gene of interest, then access the CGC for potentially available strains with corresponding transgenes.

Additional strains and information are also available at the *C. elegans* Expression Pattern website http://elegans.bcgsc.ca/perl/eprofile/index. This site is an ever-expanding collaborative effort of several labs to provide expression profiles of all worm genes. Be sure that you remind students to make note of what strain they are specifically using during each experiment throughout this course in their laboratory notebook.

Experienced worm researchers may choose to construct their own expression vector and create a new transgenic line for analysis by students in the course. An important factor to consider if you choose to make your own construct is that *C. elegans* demonstrate a phenomenon known as 'germline suppression'. In germline suppression, transgenes are inactivated in the germline and early embryo, therefore if the target gene is expressed in early embryos you will likely not be able to observe it. You may overcome this effect by using a promoter that is turned on in early embryos, known as the *pie-1* promoter (Mello *et al.*, 1992). Using this promoter to drive expression in your construct ensures that your gene can be turned on in early embryos.

For this exercise, it is important to familiarize yourself with both Wormbase and Wormatlas; both will have GFP and *lacZ* patterns for many genes, but not necessarily both for any given gene. If at any time you need help with either site, the staff of Wormbase are always eager to help and may be contacted by emailing webmaster@www.wormbase.org. Additionally, a handy tutorial book may be obtained online at http://athena.caltech.edu/~wen/userguide/index.html. You may want to consider acquiring one for each of your students.

If you would like a general explanation of fluorescent microscopy, visit http://micro.magnet.fsu.edu/primer/techniques/fluorescence/fluorhome.html.

References

Chalfie M, Tu Y, Euskirchen G, Ward WW and Prasher DC, 1994. *Science* **263**: 802–805.
Mello CC and Fire A, 1995. DNA transformation. In *Methods in Cell Biology*, **48**, Epstein H and Shakes D (eds). Academic Press: San Diego.
Mello CC, Draper BW, Krause M, Weintraub H and Priess JR, 1992. *Cell* **70**: 163–176.

3 Creation and testing of transgenic yeast for use in protein–protein interaction screening

The yeast two-hybrid system is an *in vivo* method for assaying for protein–protein interactions. This assay occurs inside living cells, which is advantageous for ensuring that proteins fold properly and assume their native conformations within the cellular environs. This method takes advantage of the way transcription factors work. These important proteins provide both the specificity *and* action required for inducing proper gene expression in cells. Some transcription factors, in particular Gal4 of yeast, accomplish this by having an intrinsic dual structural modality – a region of amino acids that specifies binding to the DNA double helix and a second domain that serves to recruit other factors required for activating transcription.

Testing for interaction between two specific proteins

There are a variety of techniques available for examining protein–protein interactions *in vivo*. The baker's yeast *Saccharomyces cerevisiae* is the most common host organism for assaying possible effectors derived from a variety of species, including humans. Yeast two-hybrid methodology is based upon the physical proximity and interaction that occurs between a DNA-binding domain and a transcriptional activation domain of

Integrated Genomics Guy A. Caldwell, Shelli N. Williams, Kim A. Caldwell
© 2006 John Wiley & Sons, Ltd

a transcription factor. The precise system that we are exploiting for the purposes of this practical experience uses the DNA-binding region of the bacterial transcription factor LexA (LexA-DB) and the transcriptional activation domain from the well-studied yeast transcription factor named Gal4 (Gal4-AD). The word 'hybrid' appropriately describes the proteins tested for interaction in this system because they represent the expression of gene fusions between the proteins to be tested and either LexA-DB or Gal4-AD (i.e. encoding hybrid proteins LexA-DB–'X' and Gal4-AD–'Y') (see Figure 3.1). When the DNA-binding domain of LexA comes into contact with the activation domain of Gal4, by virtue of an interaction between the test proteins X and Y, LexA and Gal4 join to reconstitute a functional transcription factor. This hybrid transcription factor can then initiate transcription of reporter genes that are integrated downstream (3′) of binding sites for LexA that have been engineered within the yeast chromosomal DNA.

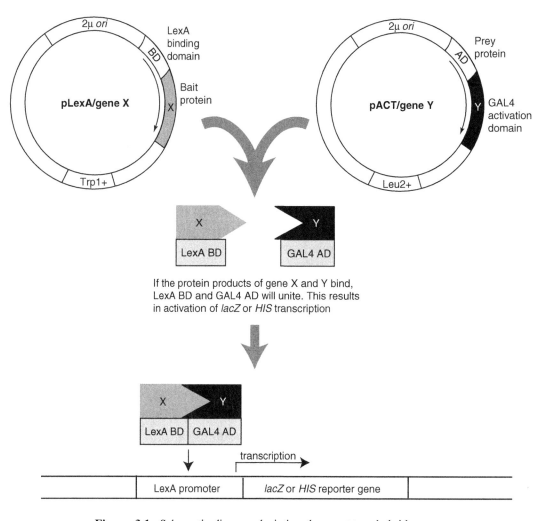

Figure 3.1 Schematic diagram depicting the yeast two-hybrid system

The system takes advantage of extensive genetic information regarding yeast nutritional, or auxotrophic, mutants. By using a strain that is deficient for the production of specific amino acids, investigators are able to select for yeasts that have taken up the experimental vectors. In our case, the *S. cerevisiae* strain L40 cannot produce tryptophan, leucine, or histidine. The LexA-DB vector (Figure 3.1; in detail at Appendix V, Figure AV.1) contains a gene that encodes the production of tryptophan (*TRP1*), whereas the Gal4-AD plasmid (Figure 3.1; in detail at Appendix V, Figure AV.2) encodes for the biosynthesis of another amino acid, leucine (*LEU2*). Selection for yeasts that have been transformed with both hybrid vectors occurs by plating yeast on growth media that is devoid of these three essential amino acids. Successfully transformed yeasts can grow on this selective media because they contain the genes necessary for producing the amino acids tryptophan and leucine. Additionally, once both vectors are harbored within a single yeast nucleus, any physical interaction between Gene X and Gene Y will bring the LexA-DB and Gal4-AD domains into physical contact. This complex can then initiate transcription of the primary reporter construct, a gene encoding for the production of histidine (*HIS3*) in yeast. Growth in the absence of histidine indicates that the gene products on these plasmids are physically interacting to induce histidine production and rescue the growth defect of the nutrient-deficient yeast.

Library screening for interaction between a specific protein and unknown proteins

In addition to providing the ability to test specific interactions between two known proteins, the two-hybrid system is amenable to screening a collection, or library, of possible interactors with any given Gene X that is fused to LexA-DB. In this scenario, the LexA-DB–X hybrid protein is typically referred to as the 'bait' that is used to fish for an interacting 'catch' protein from a library of gene products expressed as a fusion of numerous cDNAs to Gal4-AD. Briefly, libraries are constructed by isolating mRNAs from specific populations of cells. These mRNAs are then reverse transcribed to create a collection of complementary cDNAs, which are then cloned into the Gal4-AD vector. For example, a library of different Gal4-AD–cDNA vectors that represent every gene of *C. elegans* fused to Gal4-AD can be introduced into yeast cells already containing the pLexA–Gene X construct and tested for interaction. Physical interaction between the proteins encoded by Gene X and an unknown cDNA will lead to histidine production, which can be selected for by growing in the absence of histidine.

However, like all screening methods, yeast two-hybrid interactions are subject to false positives, therefore this system incorporates a *secondary* screen involving an independent color assay (Chapter 4). Physical interaction between LexA-DB and Gal4-AD not only initiates transcription of the *HIS3* gene, but also the gene encoding β-galactosidase (*lacZ*), which has been integrated downstream from the LexA-binding sites engineered into the yeast genome. β-Galactosidase (β-gal) production in response to protein interaction can be detected using a filter-based assay. In this assay, the enzyme is exposed to its colorless substrate X-gal and a cleavage reaction results in a colorimetric change. The appearance of a blue color (a positive X-gal reaction) therefore indicates physical interaction between LexA and Gal4, which can only occur if Gene X is interacting with a protein coded for by an unknown cDNA clone within the library.

Identifying interacting proteins

Following this selection scheme, the investigator must then identify what genes are responsible for detected interactions with the target Gene X (Chapter 5). Notably, yeast two-hybrid plasmids are "shuttle vectors", wherein there have origins of replication for both yeast (2μ) and *E. coli* (ori), as well as selection markers for yeast (amino acid production) and *E. coli* (antibiotic resistance). It is necessary to shuttle from yeast to bacteria due to the difficulty of separating plasmid from genomic DNA in yeast cells. Growth in the presence of the target antibiotic indicates that the plasmid has been shuttled successfully from yeast to bacteria. In this course, this transformation is accomplished by electroporation. Transformed bacteria then can be used to grow larger quantities of the library plasmid. Once this DNA has been isolated, it can be subjected to a DNA sequencing reaction, which is necessary to identify each unknown cDNA. The availability of complete genomic sequences for animals such as worms, fruit flies, mice, and humans enables investigators to identify rapidly the sequenced interacting cDNA product by genomic database searching on the Internet. Subsequent downstream applications such as functional analysis by RNA interference (RNAi) (Chapter 8) and/or expression analysis following fusion to GFP or *lacZ* can then be performed to examine the relationship of any newly identified interacting catch protein with what may be known about a specific bait protein. These experiments demonstrate the next frontier of molecular biology, because complete genome sequences only provide the raw genetic information for scientists to begin deciphering the networks of protein interactions and complexes that comprise the cellular machinery.

3.1 Small-scale transformation of *S. cerevisiae*

Transformation is the introduction of foreign DNA into a cell and is one of the most standard techniques used by molecular biologists today. Nearly all cell types, from simple bacteria and yeast to more complex mammalian cells, can be transformed. Transformation protocols are optimized for a particular cell type; within this course, students will transform both bacteria and yeast. In either case the fundamentals remain the same. The researcher places the DNA within physical proximity of the target cells, temporarily expands the membrane boundaries of the cell to enable a small portion of DNA solution to flow in, and then quickly closes the boundaries in order to trap the DNA within the cell. Cells are then allowed to recover from any damage they may have endured, and shortly begin replicating the introduced DNA. A selective pressure is then used to ensure that the cells maintain the introduced plasmid. In the case of yeast, this pressure is the absence of an essential amino acid gene that is provided by the plasmid.

You are performing the transformation detailed below in order to create the yeast strain that will be used for a protein–protein interaction screen. Just imagine, by the end of this lab you may have created an entirely new strain of yeast that has never existed! This yeast strain will contain the 'bait' vector used in the two-hybrid screen. This bait contains your *C. elegans* gene of interest (hereafter known as 'Gene X') fused to the DNA-binding domain of bacterial transcription factor LexA. Subsequent experiments will introduce different 'catch' vectors containing either known or unknown genes fused to Gal4-AD. Remember, as detailed in the yeast two-hybrid background, if the proteins produced by these plasmids interact, they will come into physical contact, thereby bringing the DNA-binding domain and transcriptional activation domain into close proximity and initiating transcription of reporter genes. Concomitantly, you will perform a second transformation with a control construct (for example, pLexA Ras) in order to create a strain for use in verifying positive control interactions during secondary screening of transformants.

Procedural overview

The procedure begins by growing an overnight culture of L40 (the yeast strain utilized in this screen) in YEPD (yeast media). Growth in this rich media ensures that you begin the transformation with a healthy liquid culture. After overnight growth, the saturated culture is diluted in fresh media and allowed to grow several more hours to ensure that the yeast culture is growing logarithmically at the time of the experiment. After this growth period you will harvest the cells by centrifugation and wash away the rich media with sterile double-distilled water. Lithium acetate is used to coat the yeast with a net positive charge (cell membranes are relatively negative due to their high phospholipid content); this positive charge will then attract the negatively charged plasmid DNA when it is added to the mixture. Salmon sperm DNA serves as a 'carrier' molecule by providing a greater negative charge when it complexes with your plasmid. This DNA complex is attracted to the positively charged lithium acetate already surrounding the yeast membrane, therefore your test DNA is in physical proximity to the yeast cells. A solution containing more lithium acetate, polyethylene glycol (PEG) and a buffer (TE) is then used to physically disrupt the yeast membrane and at the same time provide a viscous matrix that protects the cells from excess damage during vortexing. Polyethylene glycol is also a crowding agent, which means that it pushes macromolecules (such as cells and DNA complexes) into closer proximity in aqueous solutions.

After incubation in this solution, the yeast are subjected to heat shock in the presence of dimethyl sulfoxide (DMSO), which functions to enhance disruption of the cell membranes during the heat shock. The closely bound plasmid DNA should then rush into the opened membranes while the carrier DNA is too large to enter the cells. The cells, some of which now contain the test plasmid, are then harvested by centrifugation, resuspended in a buffer solution, and then plated onto selective media. Only those cells that have taken up the pLexA-DB–X or –Ras vector will be able to grow in the absence of tryptophan, due to the presence of the *TRP1* gene on this plasmid.

The transformation controls used in this procedure include two samples of yeast that are not exposed to either test plasmid. The positive control is plated onto SC complete media (containing all essential amino acids). This simply shows that the cells are healthy and were not damaged following the transformation procedure. The negative control is plated onto SC-Trp media. This shows that the culture had not undergone a mutation event that would allow the cells to grow in the absence of this essential amino acid. In other words, it confirms that the L40 strain you started with still carries a mutation that prevents it from manufacturing its own tryptophan.

Reagents (see Appendix II for recipes)

YEPD broth, sterile

Double-distilled water, sterile

0.1 M Lithium acetate (LiAc), sterile

Salmon sperm DNA, denatured ($10 \mu g \mu l^{-1}$) – available from Clontech or Sigma, stored at $-20 \,^{\circ}C$

pLexA–Gene X vector ($0.2 - 1 \mu g \mu l^{-1}$)

pLEXA–Ras vector ($0.2 - 1 \mu g \mu l^{-1}$)

0.1 M LiAc – 40 percent PEG-3350 – 1 × TE buffer

DMSO (dimethyl sulfoxide) (*Wear gloves; use in hood*)

1 × TE buffer, sterile

70 percent ethanol (for sterilizing cell spreader)

SC Complete plate

SC-Trp plates (three)

Equipment

Sterile flasks (250 ml and 500 ml)

30 °C Shaking incubator

Centrifuge tubes, 50 ml (i.e. Falcon 2098)

Tabletop centrifuge

Microcentrifuge

Microcentrifuge tubes, 1.5 ml

30 °C Waterbath

42 °C Waterbath

Bunsen burner or ethanol lamp

Sterile hood (if available)

Glass or metal cell spreader

30 °C Stationary incubator

Saccharomyces cerevisiae *yeast strain*

L40 (MATa ade2 his3 leu2 trp1 LYS2::lexA-HIS3 URA3::lexA-lacZ)

This strain is used for the LEXA-based two-hybrid system and allows for *dual selection* of protein–protein interactions using HIS^+ and β-gal reporter genes. As noted in the yeast genotype, this strain is deficient in tryptophan (*trp1*) synthesis. Details on *S. cerevisiae* nomenclature can be found at the Saccharomyces Genome Database at http://www.yeastgenome.org/.

NOTE: Yeast cultures should be treated with appropriate aseptic technique at all times or contamination will result.

Before the start of the lab

(1) Streak out L40 from a frozen glycerol stock (or other source) 5–7 days before the start of the lab onto a YEPD plate. Grow at 30 °C for 2–3 days and store at 4 °C until needed in Step 2.

(2) Inoculate 20 ml of YEPD broth in a sterile 250-ml flask with a single large colony of *S. cerevisiae* strain L40. Shake at 275 rpm overnight at 30 °C. This should be started at least 18 h before sub-culturing in Step 3.

Day of the lab (3–4 h before class)

(3) Subculture overnight cells by diluting to $OD_{600} = 0.5$ in a total of 50 ml of YEPD in a 500-ml sterile flask. This is typically achieved by pouring ∼45 ml of YEPD broth into a sterile 50-ml Falcon tube and then adding 5 ml of the overnight culture to the tube, vortexing, and checking the optical density (OD). Either add more of the culture or dilute the culture with YEPD, if needed to reach $OD_{600} = 0.5$. Pour this subculture into a sterile 500-ml flask. Grow for an additional 3–4 hours by shaking at 275 rpm and 30 °C (L40 doubles approximately every 90 min).

During the lab

(4) *Using appropriate sterile technique (see Appendix III)*, pour the culture of cells into a 50-ml Falcon tube and centrifuge (vs. a balance tube) at 2500 rpm for 4 min in a tabletop clinical centrifuge. Discard supernatant and resuspend by vortexing the pellet in 40 ml of sterile double-distilled water.

(5) Repeat centrifugation to pellet cells. Discard supernatant and resuspend each pellet in 2 ml of 0.1 M LiAc. *Do not vortex.* Gently resuspend cells by bumping tube or rocking the solution. Incubate at room temperature for 10 min.

(6) Label a single sterilized 1.5-ml microcentrifuge tube with the name of your test plasmid and another tube with 'pLexA–Ras'. For performing positive and negative control reactions, label two other control tubes C+ and C−. Dispense 10 µl of salmon sperm carrier DNA into each tube.

(7) Add 1–5 µl of the appropriate plasmid DNA (equivalent to a quantity of ∼1 − 2 µg of DNA) to each transformation tube. *Mix well* by pipeting. Be absolutely sure you have added the DNA to each tube by watching the solution enter and then leave the pipet tip. Do not add plasmid DNA to the control tubes.

(8) Add 100 μl of the yeast suspension from Step 5 to each tube (including the controls now). Mix well by gently pipeting while adding the yeast to the DNA.

(9) Add 700 μl of the 0.1 M LiAc – 40 percent PEG-3350 – 1 × TE solution to each tube. This is a viscous solution; allow it to enter and leave the pipet tip slowly as you pipet it up and dispense it. Vortex vigorously to mix (5–10 s). Incubate tubes at 30 °C for 30 min.

(10) During this incubation, label the bottom of one SC Complete and three SC-Trp plates according to the scheme described in Step 15.

(11) *Wearing gloves*, add 85 μl of DMSO to each sample. Mix well while adding. Dispose of pipet tips in a designated hazardous waste container.

(12) Heat-shock mixtures by placing tubes in a 42 °C waterbath for *exactly* 7 min.

(13) Pellet cells in a microcentrifuge using a 20-s spin at top speed (typically ∼13 000 rpm). Remove supernatant carefully using a pipettor. Add 1 ml of 1 × TE buffer to each pellet and resuspend by vortexing.

(14) Repeat rapid spin in microcentrifuge. Remove supernatant carefully. Resuspend each tube of cells in 100 μl of 1 × TE buffer by vortexing.

(15) Plate the entire mixture from each tube onto the appropriately labeled plates (Step 10). Using an ethanol-sterilized cell spreader, disperse the cells around the entire plate. Be careful that your spreader is not too hot because this will kill the yeast. Work rapidly to minimize contamination from the air. It is preferable to use a sterile hood during plating.

  (a) For pLexA–Ras, use SC-Trp.

  (b) For pLexA–Gene X, use SC-Trp.

  (c) For *positive* growth control (C+), use SC Complete.

  (d) For *negative* growth control (C−), use SC-Trp.

(16) Allow liquid to soak into agar then incubate all plates at 30 °C, inverted. Transformants will appear after ∼36–48 h.

(17) If desired, these plates can be stored at 4 °C for ∼2 months by wrapping the edges of the plates with parafilm.

Bioinformatics

Literature searching: using PubMed (www.pubmed.org) In today's computer age, keeping up with recent developments within a given field of study is both easier and more time consuming. As information expands, more time must be spent keeping abreast of new advances in the field by reading published papers as well as abstracts and information from scientific meetings. However, the Internet makes acquiring this information a more simplified task.

For many biomedical researchers, and biologists in general, one of the more frequently visited websites is PubMed, a database organized and managed by the National Center for Biotechnology Information (NCBI). PubMed contains information from hundreds of journals, including not only the basic reference data such as authors, titles and publication dates, but frequently abstracts and links to the full length papers. Like all well-managed databases, PubMed is searchable through a variety of criteria.

To familiarize yourself with PubMed, perform the following searches:

(1) Perform a search for 'two-hybrid'. Note how many total hits you receive.

(2) Perform a more specific search for '*elegans* and two-hybrid'. How does this affect your search results?

(3) Finally, try a search for reviews of the yeast two-hybrid system. To limit your search, go to the 'Limits' tab under the search box. Here you can restrict your search by publication type, language, or date.

Instructor's notes

It is the instructor's responsibility to create or obtain an appropriate yeast two-hybrid bait clone (pLexA::Gene X) for use in screening a *C. elegans* cDNA library. For information on how to clone your own gene of interest into a vector, please refer to the 'Instructor's notes' in Experiment 7 (Section 7.2, traditional restriction enzyme cloning) or Appendix I (Addgene (www.addgene.org/integrated_genomics) recombinational cloning). Addgene X can supply you with a kit containing all vectors used during the progression of this course, including both traditional and Gateway versions of pLexA and pACT2.2. This kit also contains ready-to-use bait clones if you prefer not to study your own gene of interest.

3.2 Transformation of *S. cerevisiae* to test for non-specific interaction

An important step in any scientific endeavor is the proper use of controls. Controls should be incorporated into each stage of an experiment and, if possible, even before beginning a series of studies (such as a library screen). In this exercise, you will create two control strains that will be used as controls for the protein–protein interaction screen. As discussed in the introduction to the two-hybrid system, secondary screening for potential interactors is performed using a β-galactosidase assay in yeast. Positive interactions are illustrated by the appearance of a blue color, generated when the enzyme cleaves the colorless substrate X-gal. The blue color appears when the *lacZ* gene is expressed, generating the enzyme needed to cleave X-gal. Expression of this gene is made possible when the physical proximity of the DNA binding and activation domains of the hybrid proteins activates transcription of the *lacZ* gene.

In this lab, you are creating two yeast strains by introducing a second plasmid into the L40(pLexA–Gene X) and L40(pLexA–Ras) strains created in the previous section. The pACT2.2–Raf plasmid will be transformed into these strains, creating yeast carrying two plasmids within single cells. Both Ras and Raf are known protein interactors and therefore should initiate the transcription of β-gal, which will be visualized during the next exercise. Both Ras and Raf are members of a signaling pathway that transmits signals from the surface to the nucleus of a cell. The signal may then activate a number of different effects, including cell division, differentiation, or even death. Ras is central to this signaling pathway and has been found to be defective in several types of cancer. Despite its far-reaching effects, the Ras–Raf pathway is very specific, therefore 'Gene X' theoretically should not interact with Raf, transcription of β-gal will not be activated, and no color change will occur following the X-gal assay. These strains will then be used as the positive (Ras/Raf) and negative (Gene X/Raf) controls in the subsequent secondary screening following the yeast two-hybrid screen. This control is also important because it is possible that Gene X may self-activate transcription on its own. If this is the case, it will be necessary to re-evaluate the use of a specific gene for interaction screening. It is possible that specific regions within a gene (i.e. encoding the N-terminal domain) may be sufficient for use in screening. In all cases, such a test for non-specific activation is essential prior to embarking on a screen or test for specific interaction.

Procedural overview

In this lab you begin by inoculating yeast cells into selective media (lacking tryptophan) to ensure that they maintain the original pLexA plasmid you have already introduced. A liquid culture that is allowed to grow overnight ensures that you start the procedure with healthy cells; you will again subculture to ensure that your cells are in log phase. During this subculture you will change from selective (SC-Trp) to rich (YEPD) media to ensure that the yeast are very happy and healthy at the time of transformation. Because yeast double at a rate of approximately once every 90 min, there is a minimal loss of the pLexA plasmid in this short time of growth in the absence of selection. The transformation procedure itself is the same as the previous transformation; please refer to the procedural overview in Section 3.1 for a reminder of the key points.

The ultimate goal for this lab is to introduce a second plasmid into cells already containing one plasmid, therefore at the end of the procedure you will plate the transformed cells onto selective media that lack both tryptophan and leucine. Remember, the pLexA plasmid provides the wildtype *TRP1* gene and the pACT2.2 plasmid contains the wildtype *LEU2* gene, therefore any yeast that has taken up both plasmids will be able to grow in the absence of both amino acids. These yeasts will grow somewhat slower because they have to produce their own amino acids, but they will still grow.

Reagents (see Appendix II for recipes)

SC-Trp broth, sterile

YEPD broth, sterile

Double-distilled water, sterile

0.1 M LiAc, sterile

pACT2.2–Raf (0.2–1 µg/µl)

Salmon sperm DNA, denatured (10 µg/µl) – from Clontech or Sigma, stored at −20 °C

0.1 M LiAc – 40 percent PEG-3350 – 1 × TE buffer

DMSO (dimethyl sulfoxide) (*Wear gloves; use in hood*)

1 × TE buffer, sterile

70 percent ethanol (for sterilizing cell spreader)

SC-Trp-Leu plates (four)

SC-Trp plates (two)

Equipment

Sterile flasks, 250 ml and 500 ml (two of each)

30 °C Shaking incubator

Centrifuge tubes, 50 ml (i.e. Falcon 2098)

Tabletop centrifuge

Microcentrifuge

Microcentrifuge tubes, 1.5 ml

30 °C Waterbath

42 °C Waterbath

Bunsen burner or ethanol lamp

Sterile hood, if available

Glass or metal cell spreader

30 °C Stationary incubator

Saccharomyces cerevisiae *yeast strains used*

L40(pLexA–Ras)

This strain is used for the pLexA two-hybrid system and allows for *dual selection* of protein-protein interactions using HIS^+ and β-gal reporter genes. This strain already contains the pLexA–Ras construct that you introduced in the previous laboratory experiment.

L40(pLexA–Gene X)

This strain is the same except that it contains the pLexA–Gene X construct that you introduced in the previous laboratory experiment.

   See Section 3.1 for genotype information.

NOTE: Yeast cultures should be treated with appropriate aseptic technique at all times or contamination will result.

Before the start of the lab

(1) Inoculate 20 ml of SC-Trp broth in a 50-ml sterile Falcon tube or sterile 250-ml flask with a single large colony of *S. cerevisiae* strain L40(pLexA–Ras) and another 20 ml of SC-Trp broth with a single large colony of *S. cerevisiae* strain L40(pLexA–Gene X).

(2) Shake both cultures at 275 rpm overnight at 30 °C. These should be started at least 18 h before subculturing in Step 3.

Day of the lab (3–4 h before class)

Each of the following should be performed on both of the yeast strains. Keep all tubes well labeled throughout this procedure.

(3) Subculture overnight cells by diluting to $OD_{600} = 0.5$ in a total of 50 ml of YEPD in a 500-ml sterile flask. This typically can be done by pouring $\sim$45 ml of YEPD broth into a sterile 50-ml Falcon tube and then adding 5 ml of the overnight culture to the tube, vortexing, and checking the OD. Add more of the culture or dilute the culture with YEPD, if needed to reach the $OD_{600} = 0.5$. Pour this subculture into a sterile 500-ml flask. Grow for an additional 3–4 h by shaking at 275 rpm and 30 °C.

During the lab

(4) *Using the appropriate sterile technique (see Appendix III)*, pour the culture of cells into a 50-ml Falcon tube and centrifuge at 2500 rpm for 4 min in a tabletop clinical centrifuge. Discard supernatant and resuspend by vortexing the pellet in 40 ml of sterile double-distilled water.

(5) Repeat centrifugation to pellet cells. Resuspend each pellet in 2 ml of 0.1 M LiAc. *Do not vortex*. Gently resuspend cells by bumping tube or rocking the solution. Incubate at room temperature for 10 min.

(6) Label one sterilized microcentrifuge tube 'Raf'. For performing control reactions, label two other control tubes C+ and C−. Dispense 10 µl of salmon sperm carrier DNA into each tube.

(7) Add 1–5 µl of the pACT2.2–Raf plasmid DNA (equivalent to a quantity of $\sim$1–2 µg of DNA) to the tube labeled 'Raf'. *Mix well* by pipeting. Be sure that you have added the DNA to the tube by watching the solution enter the pipet tip and then dispensing it out. Add no plasmid DNA to the control tubes.

(8) Add 100 µl of the yeast suspension from Step 5 to each tube (including the controls now). Mix well by gently pipeting while adding the yeast to the DNA.

(9) Add 700 µl of 0.1 M LiAc − 40 percent PEG-3350 − 1× TE solution to each tube. This is a viscous solution, so allow it to enter and leave the pipet tip slowly as you pipet it up and dispense it. Vortex vigorously to mix (5–10 s). Incubate tubes at 30 °C for 30 min.

(10) During this incubation, label your plates on the bottom of the plate. See Step 15 for the appropriate label for each plate.

(11) *Wearing gloves*, add 85 µl of DMSO to each sample. Mix well while adding. Dispose of pipet tips in hazardous waste container in fume hood.

(12) Heat-shock mixtures by placing tubes in a 42 °C waterbath for *exactly* 7 min.

(13) Pellet cells in a microcentrifuge with a 20-s spin. Remove supernatant carefully using a pipettor. Add 1 ml of 1 × TE buffer to each pellet and vortex to resuspend.

(14) Repeat rapid spin in microcentrifuge. Remove supernatant carefully. Resuspend each tube of cells in 100 µl of 1 × TE buffer.

(15) Plate the entire mixture from each tube onto separate labeled plates of the appropriate medium. Using an ethanol-sterilized cell spreader, spread the cells around the entire plate. Be careful that your spreader is not too hot because this will kill the yeast. Work rapidly to minimize contamination from the air. It is preferable to use a sterile hood during plating if one is available.

(a) For pLexA–Ras + pACT2.2–Raf, use SC-Trp-Leu.

(b) For pLexA–Gene X +pACT2.2–Raf, use SC-Trp-Leu.

(c) For each *positive* control (C+), use SC-Trp.

(d) For each *negative* control (C−), use SC-Trp-Leu.

(16) Allow liquid to soak into agar and then incubate all plates at 30 °C, inverted. Transformants will appear after ∼36–48 h.

(17) If desired, these plates can be stored at 4 °C, wrapped around the edge with parafilm, for approximately 2 months.

Bioinformatics

Literature searching: using the Internet Although PubMed is undeniably a useful resource, it is limited in that it contains only manuscripts that have been published. It will not contain meeting abstracts, useful information that may be found on individual researchers' websites, or more specific databases for given fields (such as Wormbase).

Fortunately, any information that has been placed on the World Wide Web (as most is these days!) can be found using an Internet search engine. Using your favorite search engine (i.e. Google, Yahoo, etc.), perform the same searches that you previously performed on PubMed. How do your results compare? A useful skill to develop is the ability to narrow the broad Internet searches. Decide what exactly it is you want to find. Search for the precise terms that you would use to describe what you want to know. Can you rephrase your request and find additional useful information? Is there a difference between the papers that you find referenced in PubMed and those from the search engine?

Find out more information about the Ras protein:

(1) What diseases are associated with Ras malfunction?

(2) What is the name of the *C. elegans* Ras gene?

3.3 Assaying for protein–protein interaction by reporter gene expression

During this exercise, you will perform a β-gal assay to test the control strains created during the previous exercise. As mentioned earlier, you expect to see a negative result from the yeast cells containing pLexA–Gene X and pACT2.2–Raf, indicating that these two proteins do not interact and were therefore not able to bring their respective DNA binding and activation domains into proximity to initiate transcription of *lacZ*. A positive result (blue color) should be observed for colonies containing pLexA–Ras and pACT2.2–Raf, because these proteins interact. Provided that this control experiment yields the expected results, it is safe to continue with the two-hybrid library screen. It is important to keep the two strains generated for this test because they will serve as the controls during the secondary screen of your putative interacting proteins. As discussed previously, if your current Gene X bait non-specifically activates transcription, your instructor will need to re-evaluate the specific region of Gene X being utilized in this experiment or the entire use of Gene X as a yeast two-hybrid bait.

Procedural overview

Please note that maintaining sterility is of the utmost importance during this lab! Yeasts cannot be selected by antibiotic resistance and their media contain sugar, thereby serving as an ideal source for bacterial contamination. Most bacteria naturally contain β-gal and would provide a false-positive X-gal result. Review 'Sterile techniques' in Appendix III before proceeding.

In this lab, you will perform a secondary screen to confirm that Gene X does not activate transcription non-specifically by testing the secondary reporter gene, *lacZ*. This procedure makes use of the ability of β-gal to cleave the colorless synthetic substrate X-gal. If it does, a blue color should appear. In the case of yeast containing a functional *lacZ* gene, the colony will gradually turn blue after exposure to X-gal, usually within 1–12 h.

The most critical steps of this procedure are Steps 7 and 8. After you have grown fresh, nicely patched colonies of your yeast strains onto a single SC-Trp-Leu plate (maintaining the selective pressure so that your cells do not lose the two plasmids), you will transfer the yeast to sterile filter paper. This provides a matrix for exposing the yeast to the X-gal, which can then be placed into a laboratory notebook as a permanent record of the assay. After this transfer you will perform the critical steps of freeze-cracking the cells. Immersing the filter in liquid nitrogen and allowing it to thaw on the benchtop will permeabilize the yeast so that the X-gal may enter the cells. When the X-gal comes into contact with the enzyme, the cleavage reaction will occur, causing the yeast patch on the filter paper to turn a bright blue color. β-Mercaptoethanol (β-ME), a reducing agent, is included in the staining solution to prevent oxidation of β-gal, which would result in a decrease in activity. The staining solution provides an optimal environment for β-gal activity. This reaction is performed at 30 °C, which is the optimal temperature for the assay. However, do keep in mind that X-gal is light sensitive, therefore incubation chambers should be kept in the dark as much as possible.

Reagents used

SC-Trp-Leu plate

10X Z buffer (prepared by combining 1 ml with 9 ml of sterile double-distilled water)

2 percent X-gal in $N'N'$-dimethylformamide (*Wear gloves!!*)

14.3 M β-mercaptoethanol (β-ME)

Sterile Whatman 1 filter papers (approximately same size as Petri dish) – may need to be cut to size before autoclaving

Liquid nitrogen in a receptacle with a wide opening (such as a Styrofoam shipping box): wear appropriate safety equipment (insulated gloves, eye protection, etc.) when using liquid nitrogen!!

Equipment

Sterile toothpicks or inoculating loops

30 °C Stationary incubator

Petri dishes

Aluminum foil

Blunt-ended forceps

Ventilated hood (if available)

Before the start of the lab (2 days)

(1) Using sterile toothpicks or inoculating loops, transfer individual colonies of yeast to a fresh plate selecting for both two-hybrid plasmid vectors (i.e. -Trp, -Leu). Create a 'patch' about the size of a small fingernail by spreading the colony across the top of the agar using the toothpick. You will plate several different types of yeast onto the same plate (i.e. L40 containing pLexA–Ras + pACT2.2–Raf as well as L40 containing pLexA–Gene X + pACT2.2–Raf). Be sure that your plates are well labeled on the bottom so that you know which patches are which (Figure 3.2A).

(2) Incubate plates at 30 °C until patches are confluent (~48 h).

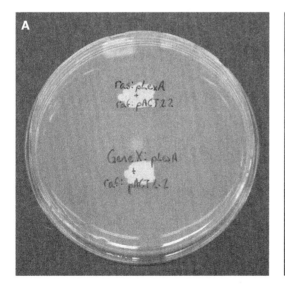 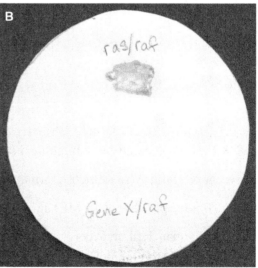

Figure 3.2 Transformed yeasts containing both bait and prey vectors, Panel A shows patches of yeast grown, then transferred to filter paper for assaying Begal activity as a readout for protein-protein interaction, as depicted in Panel B.

Day of the lab

NOTE: Gloves should be worn throughout this procedure.

(3) Prepare *lacZ* assay buffer by adding 35 μl of 14.3 M β-ME and 160 μl of 2 percent X-gal to 10 ml of 1 × Z buffer. Combine these solutions while working in a ventilated hood (if available) because β-ME has a strong odor. Mix well and cover the tube in foil.

(4) To the inverted lid of an empty Petri dish, add 2.5 ml of the freshly prepared *lacZ* assay buffer from Step 3. Place one sterile Whatman 1 filter paper disc in the center of the lid and allow it to soak up the liquid – avoid air bubbles. Cover with the bottom of the dish.

(5) Mark the orientation of another Whatman 1 filter by placing a pencil mark at the top of the filter. Do not use an ink pen! The ink will disperse throughout the filter when wet. Likewise, mark the orientation of the yeast plate (place a line at the top with a marker) that is tested.

(6) Transfer yeast colonies to the marked, dry Whatman 1 filter paper disc (Step 5) by carefully laying the filter paper on top of the yeast plate in the matching orientation. *Gently* press down on the filter paper to ensure contact with the yeast. Do not smear the paper and avoid bubbles of air in-between the paper and yeast.

(7) Using blunt-ended forceps, carefully peel the filter paper off the yeast plate and allow it to air-dry for 10–15 s.

(8) Place the filter *colony side up* in an aluminum foil 'boat' (piece of foil with edges creased upwards) and float on top of a bath of liquid nitrogen for 15 s. Using forceps, completely submerge the boat and filter in the liquid nitrogen for 5 s.

(9) Remove the boat and filter from liquid nitrogen and place the filter *colony side up* on the benchtop to thaw to room temperature (~3 min).

(10) Repeat Steps 8 and 9, *two* more times.

(11) Transfer the filter with yeast face-up onto the assay-buffer-soaked filter papers in the Petri dish lid to initiate the enzymatic reaction with X-gal. Avoid bubbles between the filters. Cover with the bottom of the dish and incubate at 30 °C, preferably in the dark.

(12) Check periodically (every 30 min) for color development. This may take 30 min to several hours. Strong interactions (i.e. Ras/Raf) yield detectable color in 60 min or less (Figure 3.2B), whereas weak interactions may take over 4 h (and can still

be valid, even if overnight). When negative controls become colored (as they will over time), all test reactions can no longer be interpreted as valid interactions. If after 4 h no color change is observed, this could be indicative of inadequate freeze/thaw to enable X-gal access to the intracellular space. Overnight incubation may yield a reaction, but the entire procedure may require repetition with an added freeze/thaw cycle.

(13) Filters should be removed from buffer and dried on top of paper towels, for several hours or overnight. Dried filters can be wrapped with plastic wrap and taped or glued into a notebook for record-keeping. Be sure to label all reactions appropriately.

Bioinformatics

Exploring the NCBI databases (http://www.ncbi.nlm.nih.gov)   The NCBI website contains many other databases in addition to PubMed. If you follow the address above, you will find the NCBI toolbar, containing a link to 'All Databases' within this resource. By clicking on this link, you will be brought to the Entrez search engine. If you performed a search here, it would query all NCBI databases. However, you can also enter specific databases by clicking on their links. Listed below are several particularly useful collections in terms of this course. Spend some time familiarizing yourself with each. It may be particularly useful to examine the tutorial section contained within each site so that you are exposed to the various functions of the database.

- *GenBank*: contains nucleotide sequence information for many genes; sequences are entered following journal publication or by direct submission of sequences.

- *Protein database*: contains amino acid sequences for proteins, submitted in the same fashion as nucleotide sequences.

- *Structure*: contains three-dimensional structure information on macromolecules. This information can be useful when searching for homology.

  (1) Using each of the above databases, search for your Gene X. Can you find the *C. elegans* version? Search by both open reading frame (for example, F53A2.4) and by gene name (i.e. *nud-1*), if available.

4   *Yeast two-hybrid screening*

4.1 Protein–protein interaction screening of a *C. elegans* cDNA library

In the previous chapter you created a *Saccharomyces cerevisiae* strain carrying your target Gene X within the pLexA plasmid. As a reminder, this plasmid carries the DNA-binding domain of LexA fused to your gene. It also carries a gene encoding tryptophan synthesis (*TRP1*), which allows for selection of yeast by their ability to grow in the absence of tryptophan, an essential amino acid. Here you will perform a yeast two-hybrid screen for gene products that interact with Gene X. Interacting proteins will be found from a population of expressed cDNAs cloned into the vector pACT2.2 to create a library that will be introduced into the *S. cerevisiae* L40 strain carrying pLexA–Gene X. The pACT2.2 plasmid contains the transcriptional activation domain of Gal4 individually fused to unknown cDNAs, as well as a marker gene encoding leucine synthesis (*LEU2*). If the product of GeneX interacts with the product of an unknown cDNA, the DNA-binding domain and transcriptional activation domain will be brought into physical proximity and initiate transcription of the yeast *HIS3* reporter gene. At the end of the transformation procedure you will spread your cells on plates lacking tryptophan (which indicates the presence of the pLexA vector), leucine (which indicates the presence of the pACT2.2 vector) *and* histidine. Growth in the absence of histidine provides for a primary screen for interaction, because it is indicative of physical interaction between the two hybrid fusion proteins.

Integrated Genomics   Guy A. Caldwell, Shelli N. Williams, Kim A. Caldwell
© 2006 John Wiley & Sons, Ltd

Procedural overview

In essence, this procedure is many small-scale transformations conducted at the same time. The reagents used in this yeast transformation are identical to those in the previous small-scale transformations of the exercises in sections 3.1 and 3.2. As in section 3.2, you will be transforming a yeast strain that already contains the pLexA–Gene X plasmid. However, there are several key differences between this protocol and the small-scale transformation. For instance, instead of transforming a single vector into the pLexA–Gene X strain, you will be transforming a *C. elegans* cDNA library. Additionally, the incubation times are greatly extended in order to increase overall efficiency of transformation. To ensure that the maximum number of cDNA clones comprising the library is screened, the number of cells used during the transformation will be increased, as will the time of the heat shock.

It is important to determine the efficiency of your transformation to be sure that enough colonies have been screened to cover adequately the size of the *C. elegans* genome. This is accomplished by performing a dilution series on one sample of the transformation mixture. These dilutions are spread onto plates lacking only tryptophan and leucine. In this manner you are selecting for yeast that have successfully taken up the pACT2.2 vector but may or may not interact with pLexA–Gene X. Formulas for determining transformation efficiency, as well as the total number of colonies screened, can be found at the end of this section. Typical yeast transformation efficiencies are in the range of 1×10^4 transformants/μg cDNA library. The transformation protocol detailed below should give sufficient transformants to identify potentially interacting clones. However, it does require manipulation of many tubes and utilizes a large quantity of relatively expensive cDNA library. Please see 'Instructor's Note' on **page 53** for scaling the transformation protocol down.

Reagents (see Appendix II for recipes)

SC-Trp broth, sterile

0.1 M LiAc, sterile

Salmon sperm carrier DNA, sheared and denatured ($\sim 10\,\mu g/\mu l$)

C. elegans cDNA library cloned into pACT2.2 vector ($\sim 1\,\mu g/\mu l$)

50 percent PEG (mol. wt. 3350), sterile

$1 \times$ PBS, sterile

70 percent ethanol (for sterilizing cell spreader)

SC-Trp-Leu-His plates (fifty)

SC-Trp-Leu plates (four)

Equipment

Sterile flasks, 250 ml and 500 ml

30 °C Shaking incubator

Tabletop centrifuge

Centrifuge tubes, 50 ml (i.e. Falcon 2098)

30 °C Stationary incubator

Microcentrifuge tubes, 1.5 ml, sterile

Microcentrifuge

42 °C Waterbath

Bunsen burner or ethanol lamp

Glass or metal cell spreader

Yeast strain used

L40(pLexA–Gene X)

This strain is used for the LexA two-hybrid system and allows for *dual selection* of protein–protein interactions using HIS^+ and β-galactosidase reporter genes. This strain already contains the pLexA–Gene X construct that you introduced in a previous laboratory exercise.

Library used

Caenorhabditis elegans cDNA library for the yeast two-hybrid system constructed in activation domain fusion vector pACT2.2, $0.5 - 1.0 \,\mu g/\mu l$, from oligo-dT primed mRNA of *C. elegans* (see Figure 4.1).

NOTE: Yeast cultures should be treated with an appropriate aseptic technique at all times or contamination will result.

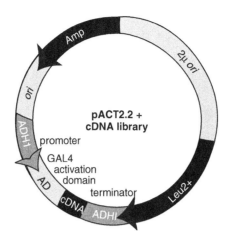

Figure 4.1   Plasmid map of the yeast two-hybrid library vector pACT2.2

Before the start of the lab

(1) Inoculate 20 ml of SC-Trp broth in a 50-ml sterile Falcon tube LexA or sterile 250-ml flask with a single large colony of *S. cerevisiae* strain L40(pLexA–Gene X). Shake culture at 275 rpm overnight at 30 °C. This should be started at least 18 h before subculturing in Step 2.

Day of the lab (∼5 h before class)

(2) Subculture overnight cells by diluting to $OD_{600} = 0.3$ in a total of 100 ml of SC-Trp broth m a 500-ml sterile flask. This typically can be done by pouring ∼95 ml of SC-Trp broth into two sterile 50-ml Falcon tubes and then adding ∼2–5 ml of the overnight culture to each tube, vortexing, and checking the OD. Add more of the culture or dilute the culture with SC-Trp broth, if needed to reach the $OD_{600} = 0.3$. Pour this subculture into a sterile 500-ml flask. Grow for an additional 5 h by shaking at 275 rpm and 30 °C.

During the lab

(3) *Using the appropriate sterile technique*, pour the culture of cells into two 50-ml Falcon tubes and centrifuge at 2500 rpm for 6 min in a tabletop clinical centrifuge. Discard supernatant and resuspend each pallet of cells in 40 ml of sterile water by vortexing.

(4) Centrifuge to pellet cells at 2500 rpm for 6 min. Discard supernatant and resuspend cells in 20 ml of sterile 0.1 M LiAc. *Do not vortex*. Gently resuspend cells well by bumping or rocking the tube.

(5) Repeat centrifugation to pellet cells. Resuspend both pellets in a total volume of 4.0 ml of 0.1 M LiAc. Resuspend both pellets by bumping or rocking the tubes and combine into a single tube.

(6) Incubate the cells for 1–2 h (longer is better, if time allows) at 30 °C with shaking at 275 rpm.

(7) Dispense 150-μl aliquots of cells into 26 1.5-ml sterile microcentrifuge tubes. Carefully add the following to each of the tubes:

   (a) 5 μl of *C. elegans* cDNA library DNA;

   (b) 1 μl of salmon sperm carrier DNA;

   (c) 400 μl of 50 percent PEG.

   Mix well by pipetting.

(8) Incubate the mixture at 30 °C without shaking for 1 h.

(9) Heat shock the mixture at 42 °C for 30 min. During this incubation, mix the tube by gentle inversion for 15 s every 5 min.

(10) Collect the cells by centrifuging for 15 s in a microcentrifuge at top speed. Carefully remove the supernatant using a pipettor and add 200 μl of 1 × PBS. Resuspend cells by gentle vortexing and pipetting.

(11) Label four sterile 1.5-ml microcentrifuge tubes: 1/10, 1/100, 1/1,000, 1/10,000. Prepare the following serial dilution series from the 26th tube:

   1/10: remove 20 μl of cells from 'undiluted' and add to 180 μl of sterile double-distilled water;

   1/100: remove 20 μl from 1/10 and add to 180 μl of sterile double-distilled water;

   1/1000: remove 20 μl from 1/100 and add to 180 μl of sterile double-distilled water;

   1/10 000: remove 20 μl from 1/1000 and add to 180 μl of sterile double-distilled water.

   Spread entire quality of each dilution onto *SC-Trp-Leu* plates (these will serve as a control to enable the determination of transformation efficiency).

(12) Spread 100 μl of each of the 25 remaining transformation mixtures onto *SC-Trp-Leu-His* plates (two plates from each transformation tube for a total of 50 plates). Allow liquid to soak into plate and incubate at 30 °C for 4–7 days to obtain transformants.

(13) Remove SC-Trp-Leu plates from incubator when discrete colonies appear (3–4 days). Determine the transformation efficiency and total colonies screened using the following formulas:

(a) *Transformation efficiency* represents how well the transformation procedure worked for each tube of yeast and library and is calculated by:

$$(\text{No. of colonies} \times \text{dilution factor})/(\text{total } \mu g \text{ library used in transformation})$$

(b) *Total colonies screened* represents the total number of colonies screened and is calculated by:

$$(\text{No. of colonies} \times \text{dilution factor}) \times \text{total number of transformations}$$

In this protocol you perform a total of 25 transformations that are used for the screen even though you plate onto 50 plates in total. Be careful not to confuse the number of transformations with the number of plates because they are different.

Caenorhabditis elegans has ~20 000 genes and a cDNA library that is representative of a collection of differentially expressed mRNAs. It is hence difficult to predict whether an individual gene is tested within such a screen, therefore a generally accepted goal for the total number of transformants screened is over one million to ensure that genes expressed at low levels are screened.

Bioinformatics

Performing BLAST searches: http://www.ncbi.nlm.nih.gov/BLAST/   In many cases scientists are interested in the sequence homology for a specific gene or protein across species. The BLAST program, available at NCBI, allows researchers to search for nucleic or amino acid sequence similarity between genes and proteins. A BLAST search is performed using an algorithm called the Basic Local Alignment Search Tool. Basically a BLAST search is a means of inputing a known sequence in an attempt to find similar sequences. These homologous sequences may be within the same species or across species lines.

Spend some time familiarizing yourself with the different types of BLAST searches that you can perform. If you are having trouble determining the differences, look at the 'Program Selection Guide' section of the website.

(1) Examine the 'Program Selection Tables' for nucleotide and protein queries, respectively.

    (a) Describe the difference between a nucleotide BLAST (blastn) versus a translated BLAST (tblastx) search.

    (b) Describe a blastp and a tblastn search.

Perform a BLAST search for homologous proteins and DNA sequences that share homology with your Gene X. You can obtain the appropriate sequence using either Wormbase or your NCBI search results.

Instructor's notes

Owing to the length of this procedure, you may want to perform Steps 3–6 before the beginning of the lab.

The *C. elegans* two hybrid library is an expensive reagent. If it is cost prohibitive for your class to purchase enough aliquots, you may wish to consider using lower quantities of it in the transformation. This *will*, however, result in significantly lower efficiencies of several clones. Additionally, it is possible to amplify CDNA libraries by growing large quantities in transformed *E. coli*. This is followed by subsequent large-scale (i.e. Maxi-prep) plasmid DNA isolation. While cost-effective, this should be avoided, if possible, as it results in significant loss of individual genes within a library during each round of amplification.

As noted in the Procedural Overview, this protocol is a rather large-scale reaction. Depending on class size and time constraints, you may wish to scale the reaction in several fashions, as outlined below:

1. *Proceed with only a single transformation as outlined, for your entire class*. After step 7, groups of students could proceed with divisions of the 26 tubes. However, keep in mind this will require calculating a single transformation efficiency and total colonies screened for the *entire* class. Also, you should be sure to discuss with students how this could affect both of these calculations (variability among different groups' techniques, etc.)

2. *Scale the transformation down for each group*. To reduce the number of tubes by half, you could proceed as outlined until step 7 and then reduce subsequent steps to 13 tubes. Alternatively, you could reduce the overnight culture and subcultures by half volumes (10 ml and 50 ml respectively).

3. *Reduce the total cell numbers*. If you choose this route, we recommend you reduce the library quantity as well. Some reductions are outlined below.

Reduction	Sub-culture volume	Resuspension Volume	Tube number	Library
One-half	Same	6 ml	Same	2.5 μl/tube
One-third	Same	9.5 ml	Same	1.6 μl/tube
One-quarter	Same	12 ml	Same	1.3 μl/tube

4. *Some combination of the above* (i.e. you could choose to both reduce the total cell numbers *and* cut the tube number in half)

4.2 Assaying for protein–protein interaction by reporter gene expression

As discussed in the introduction to Experiment 3, library screens often yield false positives, therefore it is necessary to employ a secondary screening method to validate potential interacting proteins. The initial selection for protein–protein interaction is based on growth in the absence of histidine. In much the same way that physical interaction of the hybrid proteins leads to activation of the *HIS3* gene, physical interaction between the bait and catch proteins also initiates transcription of the *lacZ* gene. The *lacZ* gene has been inserted into the yeast genome independently of the *HIS3* gene and is also downstream of LexA-binding sites.

The β-galactosidase screen is more stringent than the initial histidine selection and is used as a secondary method to eliminate false positives. The *HIS3*-dependent growth defect is 'leakier', meaning that growth may occur without true transcriptional activation. Additionally, because the yeast genome normally encodes for histidine production, it is possible that a yeast strain may revert to the wildtype state so that it can grow in the absence of histidine. Because yeasts do not normally produce β-galactosidase, it is less common to obtain a false-positive result from a *lacZ* assay.

Procedural overview

Please note that maintaining sterility is of the utmost importance during this lab! Yeast cannot be selected for by antibiotic resistance and their media contain sugar, thereby serving as an ideal source for bacterial contamination. Most bacteria naturally contain β-galactosidase and would therefore cause a positive X-gal result in any contaminated yeast cultures. Review Appendix II before proceeding.

Please see the procedural overview from the exercise in section 3.3 for a review of this laboratory protocol. However, there are a few important changes to note in this β-galactosidase assay. First, you will make an individual patch of each putative positive interactor from the two-hybrid screen onto a plate containing SC-Trp-Leu. The exact number of potential interactors will be dependent upon several factors (transformation efficiency, specific bait used) and will change for each screen, therefore it is recommended that you patch no more than 16 colonies per plate (i.e. four rows of four) (see Figure 4.2A). Although the precise number or orientation of patches is not important, it is imperative that the patches are not too close to each other. You need to discern between different yeasts and you also do not want any potential bleed-over of color from adjacent colonies on filters. At this time yeasts are patched onto SC-Trp-Leu plates to facilitate faster growth. You have already observed growth on plates lacking histidine and therefore do not need to repeat this test.

Each plate must also include a patch of yeast containing pLexA–Ras + pACT2.2–Raf and a second patch of yeast containing pLexA–Gene X + pACT2.2–Raf. These are the positive and negative controls of the reaction, respectively (see Figure 4.2B). You will judge whether the assay has been performed correctly by the appearance of blue color from the Ras/Raf patch. If this patch does not change (by overnight incubation at the latest), the assay is not valid. You will judge your experimental conditions by the lack of blue color within the Gene X/Raf patch. First, if this patch rapidly turns blue, the assay is not valid possibly due to contaminating bacteria. You can examine a wet mount of cells to confirm the presence of contaminating bacteria by microscopy. Secondly, if yeast are exposed to X-gal long enough, eventually most will turn blue. This generally takes at least 24 h, therefore when the Gene X/Raf negative control finally turns blue you must disregard any yeasts that turn after this point as being false positives for the β-galactosidase assay.

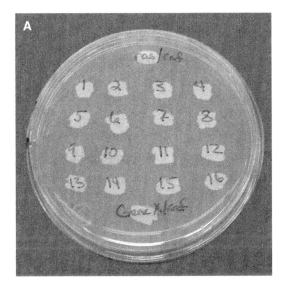 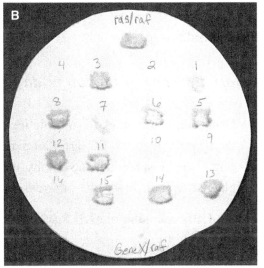

Figure 4.2 Images depicting set-up and results of yeast two-hybrid candidates of putative interactors

You will need stocks of yeast tested in the β-galactosidase assay in order to grow cultures of colonies yielding positive results, therefore colonies should be patched in duplicate onto secondary SC-Trp-Leu plates to maintain a separate stock from those utilized in the filter assay.

Reagents used

SC-Trp-Leu plates

10X Z buffer (*Note*: prepare by combining 1 ml of 10X Z buffer with 9 ml of sterile double-distilled water)

2 percent X-gal in *N'N'*-dimethylformamide (*Wear gloves*!!)

14.3 M β-ME (Sigma, catalog no. M3148)

Sterile Whatman 1 filter paper discs (approximately same size as Petri dish) – may need to be cut to size before autoclaving

Liquid nitrogen in a receptacle with a wide opening (such as a Styrofoam shipping box), ~10 cm of volume height: *wear appropriate safety equipment* (insulated gloves, eye protection, etc.) when using liquid nitrogen!!

Equipment

Sterile toothpicks or inoculating loops

30 °C Stationary incubator

Aluminum foil

Blunt-ended forceps

Before the start of the lab (2 days)

(1) Using sterile toothpicks or inoculating loops, transfer individual colonies of yeast *in duplicate to two* new SC-Trp-Leu plates. Create a 'patch' about the size of a small fingernail by spreading the colony using the toothpick onto each plate. You will plate several different types of yeast onto the same plate:

(a) L40 containing pLexA–Ras + pACT2.2–Raf;

(b) L40 containing pLexA–Gene X + pACT2.2–Raf;

(c) L40 containing pLexA–Gene X + pACT2.2–Gene Y.

Be sure that your plates are well labeled and uniform (mark orientation) for later identification. Do not label the lids of the plates, because these are easily changed by accident!

(2) Incubate plates at 30 °C until patches are confluent (~48 h).

Day of the lab

NOTE: Gloves should be warn throughout this procedure.

(3) Wrap one set of the duplicate plates in parafilm and store at 4 °C.

(4) Prepare *lacZ* assay buffer by adding 35 μl of 14.3 M β-ME and 160 μl of 2 percent X-gal to 10 ml of $1 \times Z$ buffer. Combine these solutions while working in a ventilated hood because β-ME has a strong odor. Mix well and cover tube in foil.

(5) To the lid of an empty Petri dish, add 2.5 ml of the freshly prepared *lacZ* assay buffer from Step 4. Place one sterile Whatman 1 filter paper disc in the center of the lid and allow it to soak up the liquid – avoid air bubbles. Cover with the bottom of the dish.

(6) Mark the orientation of another Whatman 1 filter by placing a pencil mark at the top of the filter. Do not use an ink pen! The ink will disperse throughout the filter when wet. Number the filter to correspond to the appropriate plate as well. Likewise, mark the orientation of the yeast plate (place a line at the top with a marker) from which colonies are to be tested.

(7) Transfer yeast colonies to the marked, dry Whatman 1 filter paper disc by carefully laying the filter paper on top of the yeast plate in the matching orientation. *Gently* press down on the filter paper to ensure contact with the yeast. Do not smear the paper and avoid bubbles of air in-between the paper and yeast.

(8) Using blunt-ended forceps, carefully peel the filter paper off the yeast plate and allow it to air-dry for 10–15 s.

(9) Place the filter *colony side up* in an aluminum foil 'boat' (piece of foil with edges creased upwards) and float on top of a bath of liquid nitrogen for 15 s. Using forceps, completely submerge the boat and filter in the liquid nitrogen for 5 s.

(10) Remove the boat and filter from liquid nitrogen and place the filter *colony side up* on the benchtop to thaw to room temperature (~3 min).

(11) Repeat Steps 9 and 10, *two more times*.

(12) Transfer the filter with yeast onto the assay-buffer-soaked filter papers, colony side up, in the Petri dish lid to initiate the enzymatic reaction with X-gal. Avoid bubbles between the filters. Cover with the bottom of the dish and incubate at 30 °C, preferably in the dark.

(13) Repeat Steps 5–12 for each SC-Trp-Leu plate containing colonies to be tested.

(14) Check periodically (every 30 min) for color development. This may take 30 min to several hours. Strong interactions such as Ras/Raf yield detectable color in 60 min or less, whereas weak interactions may take over 4 h (and can still be valid, even if overnight). It is possible, if the freeze-thaw in Steps 9–11 occurred inefficiently, that it will take longer for all color reactions to be visible. After negative controls become colored (as they will over time), any test reactions that then turn blue can no longer be interpreted as valid interactions.

(15) Filters should be dried on top of paper towels, overnight. Dried filters can be wrapped with plastic wrap and taped or glued into a notebook for record-keeping. Be sure to label all reactions appropriately.

Instructor's notes

It is possible to grow the yeast patches directly on filters; this is often the preferred method for a β-galactosidase assay because it reduces the variability caused by inefficient transfer of some cells. However, it is difficult to see the patch as you spread it on a filter and is therefore not generally used by the inexperienced. If you do decide that you wish the students to use this method, simply place an autoclaved filter onto a yeast growth plate and allow it to rest until it is saturated with liquid. Sterile toothpicks then can be used to spread the thumbnail-sized patch directly on the filter. Students need to create a duplicate plate for archival purposes to use as a source of yeast to culture from for vector isolation in Experiment.

Bioinformatics

C. elegans **Topographical Gene Expression Map: http://cmgm.stanford.edu/ ~kimlab/topomap/c._elegans_topomap.htm**   In recent years, many labs have begun performing large-scale experiments that encompass an analysis of many genes within an organism. Often the information from different screens can be found individually but rarely is it further correlated within a single given website. *Caenorhabditis elegans* researcher Stuart Kim has accrued and combined published and unpublished sets of microarray results performed throughout the research community. His database, known as the *C. elegans* Topographical Map or TopoMap, is freely available for use by anyone interested in finding genes co-expressed with their own. The database was first published in *Science* magazine (Kim *et al.* 2001. *Science* **293**: 2087–2092) and now can be found at http://cmgm.stanford.edu/~kimlab/topomap/c._elegans_topomap.htm.

(1) Review the information contained in the *Science* paper referenced above and in this site so that you have an understanding of the TopoMap program.

(2) Perform a search for genes co-expressed with your target Gene X. Look under the section entitled 'Searching and viewing the *C. elegans* gene expression terrain map'. Here you will find links to searching using a two-dimensional scatterplot or searching for genes co-expressed with your gene. Click on the link for searching for co-expressed genes. Perform each of the three basic searches offered by this site. For your search for genes localized within a specific radius of your target, start with a small radius (for example, radius = 1) but then perform two more searches of larger radii (for example, 2 and 5). If there are no co-expressed genes with your target Gene X, or if Gene X is not in the TopoMap, choose another gene so that you can learn how the TopoMap works. If your instructor does not have another suggestion for you, use *mec-2* (F14D12.4), which you are familiar with from Chapter 2.

Data from the TopoMap also can be accessed through Wormbase: In this case, go to the "Expression profile" link from the main Wormbase site. Here you can enter either a gene ranc (e.g., *tor-2*) or an open reading from designation (e.g., 37A1B.13). You may then investigate co-expressed genes within a specified radius (e.g. "1") from your target. Alternatively, you can find a gene of interest a the main query window for Wormbase, Then select the mountain listed at the 'Microarray topology map position'. At the bottom of the next page you will see a tab labeled 'Expression Space Search'. Within this area you can search for co-expressed genes within the radius of your choice.

It is important to note the key difference between information found via a microarray screen from that found from a yeast two-hybrid screen. A microarray screen suggests interaction through co-expression, but actual interaction must be shown using different techniques. A yeast two-hybrid screen shows an interaction, but if two genes are not normally co-expressed they may not interact within that organism.

5 Isolation and identification of interacting proteins

Conducting a yeast two-hybrid screen involves searching for gene products that interact with a given target or 'bait'. After conducting the screen, however, the most interesting questions remain: what cDNA clone is actually contained within each positive yeast colony and what is the role of the protein encoded by that cDNA? Answering the first question requires isolating the individual library vectors from the yeast colonies that passed both histidine and β-galactosidase selection. Subsequently one must generate a sufficient quantity of each purified vector to enable DNA sequencing of the unknown cDNA insert. These procedures are detailed within this chapter. You will then determine the identity (Chapter 6) and investigate the potential cellular role (Chapter 8) of the specific cDNAs that you have isolated.

5.1 Preparation of electrocompetent *E. coli*

To generate an easily renewable source of plasmid DNA it is necessary to move the pACT2.2 library vector from yeast to bacteria. Before you can shuttle the pACT2.2 library vector from *S. cerevisiae* to *E. coli*, it is first necessary to properly prepare the *E. coli* cells for electroporation. Electroporation is a transformation procedure that employs electrical current, rather than chemicals, to disrupt the cell membrane. In contrast to chemical transformation wherein cells are made competent for transformation by gentle treatment with ionically charged solutions, electroporation involves harsher treatment of more dense cultures. The electrical shock must be administered carefully so as to not kill all the cells while still enabling some of them to take up the plasmid DNA. This procedure details how to properly prepare and store electrocompetent cells of *E. coli* strain MH4 until you are ready to use

Integrated Genomics   Guy A. Caldwell, Shelli N. Williams, Kim A. Caldwell
© 2006 John Wiley & Sons, Ltd

them for transformation with the pACT2.2 library vector. If you do not wish to prepare your own cells, any commercially available *E. coli* strain that is *leuB⁻* deficient (such as KC8 and HB101) can be used for shuttling the pACT2.2 vector, because this plasmid enables functional complementation of leucine deficiency.

Procedural overview

The cells that will be used for electroporation must be treated very carefully. Most importantly, they must not come into contact with any charged ions. Excess ions will cause a disruption in the electrical current that passes through the cells during electroporation. Disruptions in the current will cause arcing, which will damage or kill many of the cells, therefore it is important to carefully acid-wash all receptacles used in the preparation of the cells. See Appendix II for a protocol on acid-washing labware.

Following growth in acid-washed flasks, you will carefully wash bacterial cells several times in HEPES (a non-ionic buffer solution) to remove the growth media. You will then resuspend the cells in a glycerol solution, which prevents them from rupturing when they are rapidly frozen in a dry ice – ethanol bath. This also concentrates the cells so that there will be a sufficient quantity in each sample for electroporation. The cells are frozen in this manner to preserve their overall health and can be maintained typically at $-80\,°C$ for long-term storage (several months or longer).

Reagents (see Appendix II for recipes and for acid-washing)

LB broth, sterile

HEPES buffer, 1 mM, filter sterilized

Sterile double-distilled water, stored in an acid-washed bottle

10 percent glycerol, sterile and stored in an acid-washed bottle

Dry ice

95 percent Ethanol

Equipment

Floor preparative centrifuge cooled to $4\,°C$ (i.e. Sorvall or equivalent model)

Acid-washed receptacles:

   250-ml polypropylene centrifuge bottles (two)

50-ml polypropylene centrifuge tubes (two)

Pyrex baking dish for dry ice–ethanol bath

Microcentrifuge tubes, 0.5 ml, sterile

−80 °C Freezer or liquid nitrogen storage tank

Bacterial strain used

Escherichia coli strain MH4 is a *leuB⁻* strain that can be complemented functionally by the *LEU2* gene of *S. cerevisiae* (which is present within the pACT2.2 vector).

NOTE: All cultures should be treated using an appropriate aseptic technique at all times or contamination will result.

Before the start of the lab

(1) Inoculate 5 ml of LB broth in an acid-washed 50-ml flask with a single colony of *E. coli* strain MH4. Grow overnight at 37 °C in a shaking waterbath at 225 rpm.

Day of the lab

(2) Inoculate 500 ml of LB broth in an acid-washed 2-l flask with 2.5 ml of the overnight MH4 culture. Grow to an OD_{600} of 0.5–0.6 ($\sim 2 - 3$ h).

NOTE: It is critical that the cells be kept cold at all times during subsequent steps in this procedure.

(3) Chill flask of cells on ice for 15 min. Meanwhile *pre-chill* all of the following on ice as well:

(a) 250-ml polypropylene centrifuge bottles (two);

(b) 50-ml polypropylene centrifuge tubes (two);

(c) 500-ml bottles of 1 mM HEPES buffer (two);

(d) double-distilled water, sterile (100-ml bottle);

(e) 10 percent glycerol (100-ml bottle);

(f) make sure a centrifuge and both GSA and SS-34 (or equivalent) rotors are cooled to 4 °C;

(g) place a container with ∼50–100 sterile 0.5-ml microcentrifuge tubes at −20 °C.

(4) Pour cells into the two 250-ml pre-chilled centrifuge bottles and centrifuge for 15 min at 5000 rpm in a Sorvall GSA rotor (or equivalent) that has been precooled to 4 °C in the centrifuge.

(5) Pour off supernatant and discard. *Keep cell pellets on ice.* Resuspend each pellet with 5 ml of ice-cold sterile double-distilled water. Mix by pipeting repeatedly up and down with a sterile 5-ml glass pipet with concomitant swirling. It is important not to rush this process and to make sure cells are cold by keeping them on ice while pipeting.

(6) Add 250 ml of ice-cold HEPES buffer to each cell suspension. No visible clumps of cells should be present. Centrifuge again as in Step 4.

(7) Pour off supernatant. Resuspend cells in 2 ml of ice-cold sterile double-distilled water. *Keep the cells on ice!!*

(8) Add 250 ml of ice-cold HEPES buffer to each cell suspension. No visible clumps of cells should be present. Centrifuge again as in Step 4.

(9) Pour off supernatant. Resuspend both pellets in remaining liquid. No visible clumps of cells should be present. Add 30 ml of ice-cold 10 percent glycerol to one of the resuspended pellets (keep them on ice!) and mix well. Combine this suspension with the remaining pellet. Transfer to a pre-chilled 50-ml centrifuge tube.

(10) Centrifuge in a pre-cooled SS-34 rotor (or equivalent) at 6000 rpm for 10 min (4 °C). Use a balance tube containing ∼30 ml of water.

(11) Pour off supernatant. Add 1.5 ml of ice-cold 10 percent glycerol to the cell pellet and resuspend well. Keep tube on ice.

(12) Wearing gloves, line up 0.5-ml sterile microcentrifuge tubes in an ice bucket (40–50 tubes). Prepare a dry ice – ethanol bath by filling a Pyrex glass dish ∼1/3 full with 95 percent ethanol and mixing in dry ice until the solution is viscous. Place a microfuge tube floater in the glass dish. Nearby, prepare a microfuge tube storage box for final storage at −80 °C by placing it in an ice bucket on some dry ice.

(13) Working with a partner, dispense 50 μl of cells into a microcentrifuge tube, immediately cap the tube, shake cells to the bottom of the tube with a rapid downward

flick-of-the wrist, then quickly place the tube in the float in the dry ice – ethanol bath. The cells will freeze in a few seconds. Repeat until all cells have been frozen. Remove tubes from bath, immediately wipe off ethanol with a tissue (use a fresh tissue every five tubes) and rapidly place into the storage container on dry ice.

(14) When all tubes have been frozen, transfer container to −80 °C or liquid nitrogen tank for long-term storage.

Bioinformatics

Each cDNA library created for any type of large-scale screen may contain a different proportion of a given gene, therefore two-hybrid results from different labs may reveal several different interacting partners. It is thus advisable that you search other researchers' results whenever they are available. In addition to the general information you can find on a gene in Wormbase, it is also possible to search for specific types of data. In particular for this exercise, you can limit your search to previously reported yeast two-hybrid results.

(1) Search for Gene X in Wormbase, limiting your parameters to 'Y2H interaction'.

(2) Determine if any of these interactors share the same phenotypes as Gene X by performing additional searches within Wormbase for mutant or RNAi phenotypes.

5.2 Isolation of DNA from yeast and electroporation of *E. coli*

This section details shuttling the pACT2.2 library vectors from the original yeast colonies deemed to be positive interactors into the MH4 strain of *E. coli*. This strain carries a mutation that prevents it from producing leucine (*leuB⁻*). This defect is complemented by the yeast *LEU2* gene contained on the pACT2.2 vector. Additionally, as the vector map indicates (Figure 4.1), the pACT2.2 vector also carries the *E. coli* gene encoding ampicillin resistance, therefore bacterial cells that have successfully taken up the pACT2.2 vector from a crude preparation of yeast DNA can be identified and used for amplification and further plasmid isolation. It is necessary to shuttle from yeast to bacteria because of the difficulty associated with separating plasmids from yeast genomic DNA. In bacteria, plasmids can be separated easily from bacterial genomic DNA due to the attachment of the single circular bacterial chromosome to the *E. coli* plasma membrane. The plasmid DNA remains free-floating within the bacterial extracts while chromosomal DNA can be co-precipitated with the cell membrane proteins.

Procedural overview

In this exercise you will be isolating yeast DNA (genomic as well as the introduced plasmids) and transforming MH4 bacteria (prepared in the previous exercise or purchased from a company) by electroporation with this crude extract.

DNA isolation

You begin this exercise with one culture representing each yeast colony that yielded positive results from the yeast two-hybrid tests. After these cultures grow to a high density you will precipitate the cells from the growth media by centrifugation. Cell pellets are then resuspended in a buffer (TE) and lysed using a solution containing detergents and mechanical disruption by vortexing in the presence of glass beads. Following lysis, yeast proteins are precipitated using an organic solvent composed of phenol and chloroform. When the crude solution is spun in a centrifuge, the proteins will be precipitated into a thin interphase, along with the solvent, between an aqueous solution and the glass beads. The crude DNA preparation is found within the aqueous solution and is transferred to a fresh tube. During this transfer, it is advisable to leave some of the aqueous solution behind rather than risk contaminating the final preparation with proteins from the interface.

The DNA is then precipitated using potassium acetate and 100 percent ethanol. After precipitating the DNA by centrifugation it is washed with 80 percent ethanol to remove any excess contamination, particularly salts that may have precipitated along with the DNA. This step is important because excess salts in the preparation will introduce charged ions into the electroporation mixture. You should not attempt to dislodge or resuspend the pellet; instead, gently pipette the ethanol into the tube to wash. The DNA is then resuspended in sterile double-distilled water in preparation for transformation by electroporation.

Electroporation

When you have completed the DNA isolation procedure you will have a crude preparation of DNA containing both genomic and plasmid DNA from the yeast. Both the pLexA and pACT2.2 plasmids are present within the mixture, however only the pACT2.2 plasmid contains the wildtype yeast *LEU2* gene that will complement the $leuB^-$ deficiency of the MH4 bacterial strain.

You begin this procedure by carefully cooling all the components used for electroporation. This is done to prevent excess damage to the cells because they should be treated carefully. Cells used in this procedure should be thawed gently on ice for the same reason. It is important when handling the electroporation cuvettes that you do not leave fingerprints on the metal plates of the cuvette. Fingerprints could affect the path of the current through the cuvette and result in arcing, therefore you should wear gloves at all times during this procedure.

After the cells have thawed, a small quantity of DNA from each crude yeast preparation is added to a tube of cells. Be careful that you have properly labeled everything (cuvettes, tubes of cells, and snap-cap tubes for recovery). Once the DNA has been added to each tube of cells, the mixture is added to the cuvettes. Electroporation is performed using the parameters indicated; if you observe arcing with the first few samples, lower the voltage.

Following electroporation, growth medium is added to each cuvette as a means of recovering the cells. The cells are then placed into the appropriately labeled snap-cap tubes and allowed to recover for 1 h. This enables the cells to repair any damage from the electrical shock as well as to begin replicating the pACT2.2 plasmid. This recovery/replication period is important so that the cells begin producing both leucine and β-lactamase, the protein that degrades the antibiotic ampicillin. Cells are then plated onto the selective media (M9-Leu) supplemented with ampicillin. The Leu$^+$ clones represent colonies that selectively transformed with the pACT2.2-based library plasmid *only*, because pACT2.2 contains the yeast *LEU2* gene and it can functionally complement the leucine defect inherent to *E. coli* strain MH4.

Reagents (see Appendix II for recipes)

SC-Trp-Leu broth, sterile

TE buffer, pH 8.0, sterile

Yeast lysis solution

Phenol–chloroform–isoamyl alcohol (25:24:1)

5 M Potassium acetate ($C_2H_3KO_3$)

100 percent ethanol, stored at $-20\,°C$

80 percent ethanol, stored at $-20\,°C$

Double-distilled water, sterile

M9-Leu + Amp broth (1 ml per electroporation), sterile

70 percent ethanol (for sterilizing the cell spreader)

M9-Leu + Amp plates (one plate per yeast clone)

Equipment

Snap-cap tubes (i.e. Falcon 2059)

Sterile toothpicks or inoculating loops

30 °C Shaking incubator

Microcentrifuge tubes, 1.5 ml, sterile

Glass beads (0.45–0.52 mm size)

Vortex

Microcentrifuge

Electroporation cuvettes, sterile, 0.2-cm pathlength

Electroporator (i.e. BioRad *MicroPulser*, catalog no. 165–2100)

37 °C Shaking incubator

Bunsen burner or ethanol lamp

Cell spreader

37 °C Stationary incubator

Yeast DNA isolation

Before the start of the lab

(1) Inoculate a tube containing 3 ml of SC-Trp-Leu broth for each positive two-hybrid clone identified. Use the duplicate plates made and stored in the exercise of Section 4.2 as the source of the colonies for growth. Grow at 30 °C in a shaking waterbath at 275 rpm for 24 h.

Day of the lab

Perform the following for each culture to process.

(2) Vortex the yeast culture well to resuspend all cells. Fill a 1.5-ml microcentrifuge tube with culture. Cap tightly and centrifuge for 30 s at top speed in a microcentrifuge. Discard supernatant and repeat with remaining culture. Discard supernatant again.

(3) Add 100 μl of TE buffer and resuspend the pellet by vortexing. Add 200 μl of lysis solution to the suspension.

(4) Wearing gloves, add 200 μl of phenol–chloroform–isoamyl alcohol solution (vortex this solution briefly before using to mix) and 0.3 g of glass beads. Use a Kimwipe to wipe away any beads adhering to the top of the tube because otherwise phenol will leak out.

(5) Vortex *vigorously* (two tubes at a time) for exactly 2 min.

(6) Centrifuge the mixture for 4 min at top speed in a microcentrifuge. Carefully transfer the upper aqueous phase to a new microcentrifuge tube (it is better to leave some of this behind than to take some of the interface, which will contaminate the preparation).

(7) Estimate the volume of the solution. Add 0.1 volume of 5 M potassium acetate and 2 volumes of 100 percent ethanol. Vortex to mix. Incubate at room temperature for 2 min.

(8) Centrifuge at top speed in a microcentrifuge for 5 min. Pour off supernatant.

(9) Rinse the pellet by adding 300 μl of 80 percent ethanol. Be careful to note the location of the pellet to ensure that it is not disrupted.

(10) Centrifuge again for 5 min. Dry the pellet by inverting it onto a Kimwipe and allowing it to air-dry (inverted) for 10 min. A vacuum centrifuge can also be used if available (5-min spin).

(11) Resuspend pellet in 30 μl of double-distilled water. This represents a crude solution of total DNA from yeast and includes both your pLEXA and pACT2.2 vectors, as well as all the yeast chromosomal DNA.

Electroporation of E. coli

(12) Place the following in an ice bucket (label all tubes/cuvettes):

  (a) one sterile electroporation cuvette per sample;

  (b) SOC = M9-Leu + Amp

  (c) one tube of electrocompetent MH4 cells per sample.

  Also prepare a rack of snap-cap 15-ml sterile culture tubes, one per sample, labeled appropriately.

(13) After allowing electrocompetent cells to thaw, add 2 μl of DNA from each yeast preparation to the corresponding tube of cells. Mix with the pipet tip while pipeting (swirl gently in tube).

(14) Transfer 40 µl of each cell mixture to the bottom of a chilled cuvette. Tap the cuvette on the benchtop to ensure that the solution is at the bottom. Do this for all samples prior to electroporation and return the cuvettes to ice.

(15) Wearing gloves, rapidly wipe the outer surfaces of the cuvette with a Kimwipe to remove all traces of water from the contact plates. Place the cuvette in the sample holder of the electroporation apparatus while appropriately positioning the contacts with those of the unit. Electroporate as demonstrated by the instructor at 2.5 kV voltage, 200 Ω resistance and 25 µF capacitance. Note that if arcing occurs, as evidenced by a "spark" or "pop", a completely fresh sample must be made and repeated at a lower voltage of 2.0 kV. Discard the sample that arced.

(16) Immediately remove the cuvette from the apparatus and add 1 ml of sterile ice-cold M9-Leu + Amp medium. Mix the solution up and down with the pipet to ensure that the cells have been resuspended and immediately transfer to a correspondingly labeled culture tube in a rack. Repeat Steps 15 and 16 for each sample to be electroporated.

(17) Shake culture tubes for 45 min at 180 rpm in a 37 °C shaker.

(18) Remove 200 µl of each culture and plate onto appropriately labeled M9-leucine + ampicillin agar plates.

(19) Allow the liquid to soak into the agar and then incubate all plates at 37 °C, inverted. Transformants will appear after ∼24–48 h.

Bioinformatics

Searching for othologs: OrthoGo (http://elbrus.caltech.edu/cgi-bin/igor/ortholog/ortholog)   Although aligning similar sequences from multiple species can prove useful in indicating shared function, only experimental data can show functional similarity or conservation. One way to find homologous gene products that have been shown to share functional conservation with the product of Gene X is to work through the scientific literature published on your gene of interest. Another way is by visiting another helpful bioinformatics database: the OrthoGo website.

The precursor of OrthoGo was originally developed for Wormbase curators to determine *C. elegans* orthologs and co-expressed genes in a few species. The tool subsequently was opened for public use to show orthology and returns hits displayed using the annotations of the Gene Ontology Consortium. The GO project is a group effort to fulfill the need of the scientific community for consistent descriptions across multiple databases, therefore the OrthoGo program will return results for gene products that display similar function, even if those functions are described differently.

(1) Run a search for your Gene X in the OrthoGo database. Do the results you receive correlate with the sequence homologs and conservations that you found using BLAST and Clustal?

(2) Do the GO annotations (molecular function, biological process, and cellular component) described in other species give you any indication of possible RNAi phenotypes that you might observe when studying interacting gene products of Gene X? Take note of these results and use them in Experiment 8 when deciding which RNAi assays to perform.

5.3 Small-scale isolation of plasmid DNA from *E. coli*: the mini-prep

After you have completed the exercise in Section 5.2 and allowed time for transformants to grow, the next step is to purify the pACT2.2 library plasmid from the bacterial genomic DNA. This purification is important because you will use the purified DNA in a DNA sequencing reaction (next section) that should yield the information used to identify each potentially interacting unknown cDNA contained within the individual library plasmids (Exercise 6).

Procedural overview

This procedure begins by growing overnight cultures of individual colonies representing successfully transformed MH4 cells that have taken up the pACT2.2 library plasmids. As a reminder, this vector provides the wildtype yeast *LEU2* gene (rescues the *leuB⁻* defect) and the β-lactamase that degrades ampicillin. After growing overnight cultures, you will proceed with plasmid isolation using a commercially available kit. Although it is possible to purify plasmids using basic laboratory techniques, the advent of inexpensive commercially available kits has greatly simplified this procedure. We recommend and outline a procedure utilizing the Qiagen QIAprep Spin Column Kit. It is at your instructor's discretion to use this kit. If you are provided with a different spin column kit used for isolating bacterially grown plasmids, follow the protocol provided with the kit. The plasmid DNA that is ultimately eluted from the columns represents a highly pure sample that can be used immediately for many downstream applications (no need for further purification), such as DNA sequencing. Here, the buffers utilized in the Qiagen kit are described. Other kits use similar methods with different names.

Cells in the overnight cultures are precipitated by centrifugation at the beginning of the mini-prep procedure. After centrifuging, the cell pellets are resuspended in Buffer P1, which acts not only as a gentle resuspension solution but also contains RNase A, which degrades RNA that otherwise would contaminate the final DNA sample. Resuspended cells are then placed in Buffer P2, which is an alkaline solution that contains detergents (i.e., SDS) that lyse open the cells and begin to digest the bacterial proteins. Cells therefore should be exposed to P2 only for several minutes and for no more than a maximum of 5 min. Extended exposures can result in degraded DNA samples because the released nucleases from bacterial cells begin to act on the now exposed nucleic acids. Buffer P2 is inactivated by Buffer N3, which in addition to being a neutralization solution also adjusts the solution to the high-salt condition that facilitates DNA binding to the filter-membrane.

After neutralizing Buffer P2 using Buffer N3, the proteins and cellular debris are pelleted by centrifugation. Because the bacterial genome is bound to the cell membrane, it is also pelleted with the debris. The aqueous solution that results from this centrifugation contains the free-floating plasmid DNA, although it is not yet completely purified. The aqueous solution is then added to a specially designed spin column containing a silica-based resin that will bind to plasmid DNA. It is important when pouring the aqueous solution into the column that you do not allow any of the precipitated debris to fall into the column, because it will clog the filter and reduce the yield of plasmid DNA. The aqueous solution is then washed through the column by centrifugation. The DNA becomes bound to the specially designed silica filter whereas contaminants such as proteins and any remaining RNA wash through. The filter-bound DNA is then rinsed with Buffer PB, which removes the excess nucleases that are found within many bacterial strains. The column is then rinsed with Buffer PE (composed mainly of ethanol), which serves to further clean the filter of potential contaminants and removes much of the salt used to bind the DNA to the filter. After the PE rinse, it is necessary to centrifuge the column a second time to dry the filter. If your lab is located in a particularly humid area, increase the spin time to 2 min. This removes residual ethanol that otherwise could contaminate the final preparation. After centrifuging the column to dry it, Buffer EB (elution buffer) is added directly to the column. This aqueous buffer has a relatively high pH compared with the preceding solutions; it is this difference that enables elution of the DNA from the silica. After incubating the column in EB, the flow-through from the final centrifugation contains purified plasmid DNA that is ready for most downstream applications.

Reagents (see Appendix II for recipes)

LB broth + 100 μg/ml ampicillin

Qiagen QIAprep Spin Column Kit

Equipment

Snap-cap tubes (such as Falcon 2059), sterile

Toothpicks or inoculating loop, sterile

Microcentrifuge tubes, 1.5 ml, sterile

Microcentrifuge

Vortex

Vacuum aspirator, optional

37 °C shaking waterbath

Before the start of the lab (Day 1, 16–19 h before the start of class)

(1) Inoculate 3 ml of LB + 100 μg/ml ampicillin with one colony from each MH4 transformation into a 15-ml snap-cap tube, labeled appropriately.

(2) Place in a 37 °C waterbath and shake for ∼14–16 h at 225 rpm.

Day of the lab (Day 2)

(3) Label the following set of tubes with the name of each culture to isolate plasmid from:

  (a) 1.5-ml microcentrifuge tube with lid;

  (b) 2-ml collection tube – place the QIAprep column from the Qiagen kit in the 2-ml tube;

  (c) 1.5-ml microcentrifuge tube with lid cut off (note: this tube should be wide enough to accommodate the spin column).

(4) Wearing gloves, lightly vortex the culture and carefully pour ∼1.5 ml of culture into an appropriately labeled 1.5-ml tube (with lid). At this time, place Buffer P1 (at 4 °C) on ice for use below.

(5) Spin down the bacteria in the microcentrifuge for 1 min at top speed (typically 13 000 rpm).

(6) Pour off the supernatant or remove with a vacuum aspirator. Be careful not to suck up the bacterial pellet if using an aspirator.

(7) Repeat Steps 4–6. Remove as much supernatant as possible.

(8) Completely resuspend pelleted bacterial cells in 250 μl of Buffer P1 by vortexing.

(9) Add 250 μl of Buffer P2 and gently invert the tube *eight times* to mix. *Do not vortex*! Work quickly because the cells should not be in Buffer P2 for more than 5 min!

(10) Add 350 μl of Buffer N3 and invert the tube immediately, but gently, *ten times*. *Do not vortex*!

(11) Centrifuge for 10 min at top speed in a microcentrifuge.

(12) Carefully pour the supernatant from Step 11 into the QIAprep column that is resting inside the 2-ml collection tube. Be careful not to pour any of the debris into the column.

(13) Centrifuge for 1 min at top speed in a microcentrifuge. Discard the flow-through, but save the collection tube.

(14) Wash the QIAprep spin column by adding 500 μl of Buffer PB.

(15) Centrifuge for 1 min and then discard the flow-through, but save the collection tube.

(16) Wash the QIAprep spin column by adding 750 μl of Buffer PE.

(17) Centrifuge for 1 min at top speed in a microcentrifuge. Discard the flow-through but save the collection tube.

(18) Centrifuge the column once more as above for 1–2 min at top speed to remove residual wash buffer.

(19) Place the QIAprep column in the 1.5-ml centrifuge tube that does not have a lid (the 2-ml tube can be discarded).

(20) Elute the DNA by adding 50 μl of Buffer EB to the center of the QIAprep column. Watch carefully to be sure that the solution covers the silica and does not cling to the sides of the column. Leave to stand for 1 min.

(21) Centrifuge for 1 min at top speed in a microcentrifuge.

(22) Discard the QIAprep column and transfer the DNA to a fresh 1.5-ml microcentrifuge tube with a lid. Label your tube appropriately. Store purified plasmid DNA at −20 °C.

Instructor's Note

It may be unnecessary to cut the lids off of the final set of collection tubes if the microcentrifuge you use can accomodate the number and size of open-lid microcentrifuge tubes and spin-columns.

Bioinformatics

Analyzing homology by aligning sequences:
Clustal (http://searchlauncher.bcm. tmc.edu/multi-align/multi-align.html) By now
you have some experience with BLAST searching to determine homology. A BLAST
search will query your entry against the databases selected and return information
showing portions of your sequence aligned with similar regions of sequence from a
single species. However, it is often useful to examine homologous sequences from
multiple species in an attempt to find conserved domains. Conserved domains often can
be helpful in determining if sequences function in a similar capacity across species or
simply share multiple regions of homology. Even if overall homology between sequences
is low, conservation of particular regions (domains) between sets of sequences may be
functionally significant.

One way to determine the conservation of sequence across several species (the best
way to observe conserved functional domains) is by aligning multiple sequences at once.
The Clustal program is an online resource for performing just such an alignment.

(1) Obtain the protein sequence of Gene X and homologous sequences from other
species (perform a BLAST search as outlined in Chapter 3 if you do not already
know Gene X's homologs).

(2) Input all your sequences into ClustalW and perform the alignment; if you wish, you
can change the name of each to the appropriate species. When inputing sequences,
each individual sequence should be preceded by '>', followed by '//'. For example,
>Gene X atg. . . caa//. It is sometimes useful to compile all sequences in a word
processing file initially before pasting all of them into the ClustalW window.

Your results will be displayed with the sequences arranged so that conserved amino
acids are aligned on the page. At this point in time, you can continue with one of two
options:

(a) Use a highlighter to illustrate all identical sequences. If you are a real biochemical
wiz and know the properties of all the amino acids, you can also indicate which are
functionally similar to each other using a different color.

Alternatively:

(b) Towards the bottom of the page, the sequences will be displayed aligned in a green
box. Copy the information in this box and click on the link to the 'BOXSHADE'
server. Choose an appropriate output format and run the BOXSHADE program. It
is generally advised to use the 'PICT' format but 'RTF' will display results in color.

Another outstanding Internet-associated resource for performing multiple sequence alignments can be found at the ExPASy (Export Protein Analysis System) website: http:://US.expasy.org/

This site, hosted by the Swiss Institute of Bioinformatics is an outstanding resource for sequence and structural analysis of proteins. Likewise, the ClustalW interface at the European Bioinformatics Institute (EBI) enables one to generate alternate, and colored, outputs for multiple sequence alignments: http://www.ebi.ac.uk/clustalw

5.4 Sequencing of two-hybrid library plasmid DNA vectors

After purifying plasmid DNA using the mini-prep procedure of the preceding exercise, you are ready to proceed with the sequencing of each vector. This procedure requires several key reagents and equipment. Please review the protocol and instruction manuals carefully before proceeding.

It is also important that you know the quantity of DNA from each mini-prep purification reaction (see Instructor's notes). Quantitate each sample using a DNA spectrophotometer (follow protocol in instruction manual) or ethidium bromide fluorescence.

Procedural overview

Cycle sequencing reactions work using the same basic principles as any polymerase chain reaction (PCR). A DNA polymerase creates a complementary sequence to the template strand by incorporating new nucleotides into a chain growing from a primer (short starting sequence) (Figure 5.1). In the case of a cycle sequencing reaction mixture, most of the nucleotides are the typical deoxynucleotides (dNTPs). However, the mixture also contains dideoxynucleotides (ddNTPs). These nucleotides lack the hydroxyl group at position 3 on the phosphate ring (see Figure 5.2). Each type of ddNTP (A, T, G, or C) is labeled with a different fluorescent molecule. When the DNA polymerase incorporates

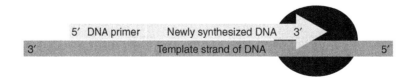

Figure 5.1 Priming and extension of a new DNA strand from a DNA template by DNA polymerase

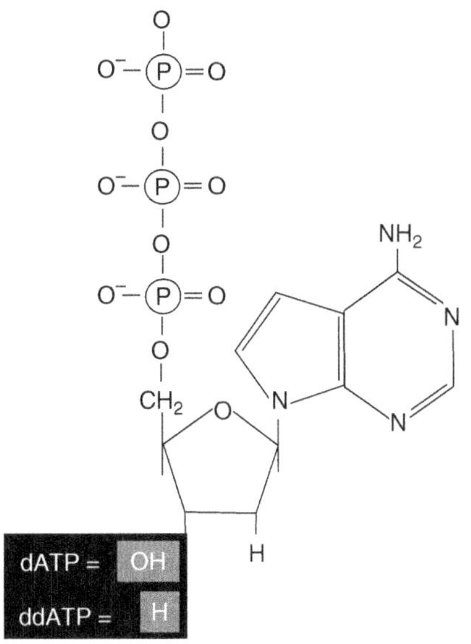

Figure 5.2   Representation of an adenine nucleotide: dioxy vs. dideoxy

a ddNTP into the chain, further extension is halted because new bases cannot be added in the absence of the hydroxyl group, therefore through random incorporation of ddNTPs amplification of the template strand theoretically should be halted at every base within the sequence. An automated sequencer then reads the fluorescent signals and digitally reassembles the sequence.

The other major deviation from a normal PCR reaction is that a cycle sequencing reaction utilizes only a single primer for strand extension. In this way, amplification can proceed only in one direction, thereby enabling the automated sequencer to reassemble the sequence in a sequential manner.

In this laboratory exercise, you will sequence the unknown gene insert(s) contained within the pACT2.2 plasmid. The reaction is primed using a sequence complementary to the DNA immediately upstream of the site of gene insertion, therefore polymerase extension will move through the unknown insert. By far the most common source of cycle sequencing reaction mixtures is Applied Biosystems, Inc. This protocol is based on the use of ABI's BigDye® Terminator v3.1 Cycle Sequencing Kit. The BigDye terminator reaction mix already contains the polymerase, dNTPs and labeled ddNTPs needed for the reaction. The polymerase used in this reaction is a *Taq* polymerase, which is a thermostable enzyme. Use of a thermostable enzyme enables multiple cycles without the need to add more polymerase after each denaturing step. You must supply the plasmid DNA (isolated in the previous exercise) and the template-specific primer. After completing the cycle sequencing reaction, it is necessary to remove all components from the mixture reaction except the final sequencing products generated. This is done by using specially designed clean-up columns. These columns work by separating molecules based on molecular weight. Large molecules, such as DNA, will pass through the columns whereas smaller molecules, such as unincorporated nucleotides, will be bound in the column. After the reactions have been cleaned, samples should be desiccated completely using a vacuum centrifuge. Samples are then ready for analysis at any facility that possesses an automated DNA sequencing machine.

Reagents

- pACT2.2- vectors isolated from MH4 with unknown gene inserts (of known concentration)
- pACT2.2-5 primer (1 pmol/μl)
    pACT2.2-5: 5′-ATACCACTACAATGGATGATG-3′

- ABI Prism BigDye Terminator Cycle Sequencing Reaction Mix*:
    A – dideoxy terminator dye labeled with dichloro[R6G]
    C – dideoxy terminator dye labeled with dichloro[ROX]
    G – dideoxy terminator dye labeled with dichloro[R110]
    T – dideoxy terminator dye labeled with dichloro[TAMRA]
    AmpliTaq DNA polymerase
    Deoxynucleotide triphosphates (dATP, dCTP, dITP, dUTP)
    $MgCl_2$
    Tris-HCl buffer, pH 8.0

(* To prevent degradation of the BigDye components, it is recommended that you aliquot it into single-use 0.2-ml PCR tubes (8 μl into each tube) upon arrival. This prevents freeze – thaw associated degradation.)

- Centri-Sep column (Princeton Separations, Inc., catalog no. CS-100) (or equivalent columns for cleaning sequencing reactions)

Equipment

Thin-walled PCR tubes, 0.2 ml

Thermocycler

Microcentrifuge

Vacuum centrifuge

NOTE: *Automated sequencing can be performed at many commercial facilities for a fee. A basic Internet search will yield several options if your institute does not already have access to a sequencing facility. Costs for this process are dropping continually.*

Cycling parameters for DNA sequencing

The following parameters should be programmed into a thermocycler prior to setting up the sequencing reaction:

(1) Quantitate each sample using a DNA spectrophotometer. Because spectrophotometers vary between manufacturers, it is best to follow the instructions for the specific device available to your lab. In all cases, DNA concentration can be measured at

Step	Temp.	Time	Purpose of each step
1	96 °C	0:10 min	Denaturation of DNA
2	50 °C	0:05 min	Annealing of primers
3	60 °C	4:00 min	Extension by polymerase
	Go to Step 1, 25×		Cycling
4	4 °C	Indefinite	Keeps sample cool until use; not necessary if you will be removing the samples as soon as the reaction is complete

Thermal 'ramping time' should be set to $1\,°C\,s^{-1}$ between each step. It is important always to use a heated lid, if available, to prevent condensation of the PCR reaction into the lid of the tube. Otherwise, each reaction must be overlaid with a layer of PCR-grade mineral oil.

absorbance A_{260}. An A_{260} value of 1.0 is equal to $50\,\mu g\,ml^{-1}$ of double-stranded DNA. Commonly, the purity of a DNA sample can be judged by examining the ratio of A_{260} to A_{280}, where absorbance at A_{280} is a measure of residual protein contamination. A ratio of 1.8 is pure DNA. Any ratio between 1.3 and 2.0 should be of sufficient purity to proceed.

(2) Calculate the volume of plasmid DNA template required to obtain 300 ng for each sequencing reaction. Optimally, you wish to use $<8\,\mu l$.

(3) Each reaction will contain the following (keep all solutions on ice). To the $8.0\,\mu l$ of terminator reaction mix (in a 0.2-ml tube), add:

Primer (1 pmol/μl)   3.2 μl
300 ng of template DNA $x\,\mu l$
Double-distilled water $20-(11.2+x)\mu l$ final volume

(4) Mix *well* using a pipetor and spin briefly in a microcentrifuge.

(5) Place tubes in a preheated thermocycler. Use the preset program – your instructor will demonstrate the use of this machine. See the cycling conditions outlined for this program above. Reactions take ~2 h. Once reaction has begun, proceed with Step 6.

(6) Obtain one Centri-Sep spin column for each sequencing reaction performed. Gently tap the column to cause the gel material to settle to the bottom of the column. If not using Centri-Sep spin columns, proceed with the manufacturer's instructions.

(7) Remove the upper end cap and add 800 μl of double-distilled water to the column to rehydrate the gel.

(8) Replace the upper end cap and then invert and bump the column a few times to mix the water and gel material. Avoid forming bubbles in the gel. If bubbles do form, tap the column to release them. Allow the gel to hydrate at room temperature for at least 2 h.

(9) Remove any air bubbles by tapping the column and allowing the gel to settle.

(10) Remove the upper end cap first and then remove the bottom cap. Allow the column to drain completely by gravity into a sink or waste container.

(11) Insert the column into the 2-ml wash tube provided.

(12) Spin the column in a microcentrifuge at 3000 rpm for 2 min to remove interstitial fluid.

NOTE: It is important to place the column into the microfuge with the tube notch facing outward everytime you spin.

(13) Remove the column from the wash tube and insert the column into the 1.5-ml microcentrifuge tube.

(14) When the PCR reaction is complete, remove the sequencing reaction mixture from its 0.2-ml cycling tube and load the entire volume carefully onto the center of the gel material.

(15) Place your column in the microcentrifuge, with the plastic notch on the column facing outward. Spin the column for 2 min at 3000 rpm.

(16) Your sample is located in the 1.5-ml collection tube – discard the column.

(17) Dry the sample in a vacuum centrifuge for ~30 min. Some sequencing facilities will perform this step for you. The sample is now ready for processing by an automated DNA sequencing machine (i.e. Applied Biosystems Inc., Prism Model 310 or 377).

Instructor's notes

Quantitating DNA

The instructor may wish students to learn DNA quantitation methods. When quantitating DNA with a spectrophotometer the sample is placed within a special cuvette that fits into the chamber of the machine. Light enters through one end of the cuvette and a sensor on the other side measures how much light is able to pass through the sample. This amount is the absorbance. Many spectrophotometer models today use quartz microcuvettes to hold the sample during quantification. These cuvettes require less total volume than older models, minimizing sample loss during quantification. Furthermore, disposable microcuvettes are also available from several manufacturers; these also require a low total volume. To further reduce sample loss, we recommend diluting a small amount of your sample to use for quantification. For example, if your cuvette will hold a total volume of $100\,\mu l$, add $2\,\mu l$ of your sample to $98\,\mu l$ of the elution or resuspension buffer in a fresh tube. In the case of a spin column clean-up, samples are eluted with Buffer EB. Thus, use EB for your dilution and for blanking the spectophotometer. After mixing this solution well, use the $100\,\mu l$ to quantitate your DNA. If your spectrophotometer can calculate dilution factors, simply enter a factor of '50' (total volume of $100\,\mu l$ divided by $2\,\mu l$ of sample). If your machine cannot calculate dilution factors, simply multiply the concentration measured by the machine by 50. This will enable you to obtain an accurate measure of DNA concentration without excess sample loss. In the rare case that your machine does not calculate DNA concentration, simply calculate the concentration by hand using the formula above and then multiply by your dilution factor. In all cases, $A_{260} = 1.0$ is equivalent to $50\,\mu g/ml$ of double-stranded DNA.

It is also possible to quantitate DNA by running it on an agarose gel along with markers of known concentrations (for instance, the molecular marker recommended in this text has bands of specified concentration). Simply prepare your samples as detailed in the exercise in Section 7.3, using a molecular weight ladder with bands of known concentrations. However, we do not recommend this more imprecise method when attempting to quantitate samples that will be used in subsequent sequencing reactions.

The most common cause of DNA sequencing failure is an impure template. If you use something other than the recommended kit from Qiagen for plasmid purification and do not obtain sequence information, we suggest that you repeat plasmid isolation using this kit.

6 Using bioinformatics in modern science

Advances in computer science have provided molecular biologists with a wealth of tools with which to extend experimental data into digital information. The burgeoning field of bioinformatics is representative of this interface. By this time you should have obtained the electronic information from your sequencing reactions. This chapter details how you should deal with this information, from reading the files, to identifying your interacting products, and, finally, choosing RNAi targets. Carefully read over each of the sections below, detailing the different steps you will take in analyzing your sequence results. At the end of the explanation sections you will find a list of all the pieces of information you should gather to facilitate your choice of an RNAi target with which to pursue subsequent phenotypic analysis. You will also find an example cDNA that you can use to practice some of the steps before beginning your own analysis.

Furthermore, you will have several options for obtaining the required information. In all cases, you can individually visit each of the sites previously discussed in the bioinformatics exercises. However, Wormbase also contains many of the tools and information you will need and can therefore be used as a kind of 'one-stop shopping' for your bioinformatics needs.

The last portion of this chapter contains different questions testing your bioinformatics knowledge. If your instructor does not require you to complete these questions, we highly recommend that you take the time to work through them on your own. Experience makes everything easier; using bioinformatics databases is no different. The field of bioinformatics is dynamic and rapidly growing. There is no substitute for just getting online and 'playing'.

Integrated Genomics   Guy A. Caldwell, Shelli N. Williams, Kim A. Caldwell
© 2006 John Wiley & Sons, Ltd

6.1 DNA sequence chromatogram

Your sequencing results from the exercise in section 5.4 were likely to have been provided to you in a chromatogram format (also called an electropherogram; see Figure 6.1) and possibly a text file. If you do not already have access to a program for opening these files, you will need to download one. Several versions are available free of charge:

(1) ABIView (PC) (http://bioinformatics.weizmann.ac.il/software/abiview/abiview.html) is a freeware beta program that you can download for opening your chromatograms.

(2) Chromas (PC) (http://www.technelysium.com.au/chromas.html) is a freeware program but you may not have access to all its capabilities without paying a small fee. If you choose to use Chromas to read your files, it will not automatically open emailed files. You will need to open them manually from within the program itself. One useful feature of Chromas is that you can BLAST your sequence information from directly inside the program. You can even choose to limit your search parameters by species, which will greatly facilitate identification of the sequenced cDNA.

(3) FinchTV (Mac or PC) (http://www.geospiza.com/finchtv/) is similar to the previous program in that you may perform BLASTs from within the program. It has an

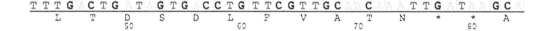

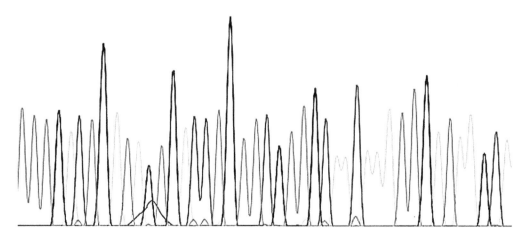

Figure 6.1 Example of a typical chromatographic read-out following automated DNA sequencing (Note that a color version of this figure can be viewed on the book's website)

additional advantage – the ability to display the entire chromatogram in a single plane. This will facilitate printing for documentation within your notebook.

(4) EditView (Mac) (http://www.appliedbiosystems.com/support/software/dnaseq/installs.cfm; http://www.appliedbiosystems.com/support/software/3100/conversion.cfm) is a freeware program that many Mac users may choose for analyzing their sequencing results. You may also need to download an additional conversion utility because most automated sequencers use PCs.

(5) 4 Peaks (Mac) (http://www.mekentosj.com/4peaks/) is another Macintosh program that allows users to view sequencing files. This program is also capable of BLASTing sequencing results and works with most automated sequencing formats, alleviating the need for conversion of PC-based files.

You may also find additional programs by performing an Internet search for sequence analysis software or tools. However, for the purposes of this text, any of the above programs are sufficient. When you examine the chromatogram, you will see peaks of one of four possible colors. Generally the colors are: red for thymine (T), green for adenine (A), blue for cytosine (C), and black for guanine (G). The height of the peak indicates the strength of the signal. Usually the display will show the chromatogram along the bottom of the window; above the peaks will be the one-letter designation for that nucleotide. The computer programs are typically very accurate at reading the information but you will often find 'N' within your sequence. An 'N' indicates that the program was unable to assign a nucleotide identity to that peak.

Failure to assign a nucleotide to a peak can indicate several things: the sequencing reaction itself may have been of poor quality in that region, the signal may not be strong enough to allow for accurate designation or may be overwhelmed by a particularly strong adjacent signal; and there may have been many bases of the same identity within that stretch of sequence and the machine is unable to count them individually. If you do find an 'N' within your sequence, manually examine the chromatogram. Often you may be able to make a decision based on the peaks displayed.

6.2 BLASTing your sequence

For this exercise, your goal is to assign a gene identity to each sequence. This is easily managed by performing a BLAST search (see Section 4.1) of your sequencing data. If your sequence analysis program cannot directly perform a BLAST search, you will need to perform one of your own to determine what the sequence actually represents. To do this, simply highlight and copy the text information from your sequencing data and paste it into a BLAST program. If your program will perform BLAST, you need only choose that option. Provided that your information is relatively 'clean' (few undeterminable bases), the first result of your BLAST search should represent the *C. elegans* gene corresponding to your sequence information.

Your BLAST results will probably contain several pieces of information. A very clean read of sufficient length (150–200 bases) should return the open reading frame corresponding to a given cDNA that was represented in your *C. elegans* two-hybrid library. Alternatively, your BLAST search may return genomic information corresponding to a specific cosmid sequence. A cosmid is like a very large plasmid and they are frequently used by scientists to subclone large pieces of genomic DNA. Cosmids representing the entire genome of an organism can then be sequenced and used for reassembling the genomic information for that organism. Often, in addition to submitting individual gene sequences (genomic and cDNA), researchers will also submit the cosmid information to databases. For example F53A2.4 (the open reading frame number) corresponds to the *C. elegans* gene *nud-1*, the fourth gene on cosmid F53A2. When BLASTing sequencing results, you will often see results indicating base pair numbers specifying a region within a cosmid (i.e. *nud-1* is found between bases 23418 and 24427 of cosmid F53A2). If this is the case, you need simply to obtain the cosmid information (available in the GenBank database of NCBI) in order to look up the appropriate region. At the beginning of the cosmid text information, preceding the actual sequence information, you will find a list of all genes contained within that cosmid and which bases they span. Use this information to determine which open reading frame corresponds to your BLAST results.

It is important when BLASTing your information that you identify your sequenced products to the level of an open reading frame. This information will be essential for determining which gene you will use as an RNAi target in subsequent experiments in this course.

6.3 Evaluating sequence results and choosing an RNAi target

During the course of your yeast two-hybrid screen, you performed two different levels of controls. First, you tested for transcriptional activation of the *HIS3* gene, which results in growth in the absence of histidine. Secondly, you tested your yeast directly for transcriptional activation of the *lacZ* reporter, which yields a positive color reaction when assaying for the *lacZ* gene product β-galactosidase. Neither of these results should occur without the DNA binding and activation domains of the hybrid proteins being tested coming into physical proximity. However, there are still false positives commonly found by two-hybrid screening. Some gene products are able to activate transcription on their own (i.e. transcription factors and histones) and other genes may show ubiquitous binding in the artificial environment of the two-hybrid system.

So now that you have confirmed the identity of the interacting gene products that you isolated from your protein–protein interaction screen, it is necessary to evaluate your results before beginning further analysis. As an initial step to discerning any functional information on your positive hits, it is advisable to take advantage of the mature (and evergrowing) bioinformatics information that exists for *C. elegans*. We recommend that you search for each of your sequenced clones in both the *C. elegans* Gene Expression TopoMap (which provides gene co-expression information) and also the Interactome datasets. Co-expressed genes often work in similar pathways (but not always); likewise, the Interactome information will provide you with data indicating any gene product interactions already known for your clones. You may find that one of your clones is known to interact with another gene product, which in turn interacts with the product of Gene X (or you may validate a previously identified direct interaction between the clone and Gene X). All of this information will be used in deciding which gene you will choose to analyze by RNAi.

For each clone sequenced, record the following pieces of information in your notebook, if they are available. Also record the information for your Gene X to facilitate comparison. You may want to place the information in a grid or report sheet to facilitate this comparison.

- Open reading frame

- Gene name

- TopoMap Expression Mountain (name of mountain, not just number)

- Yeast two-hybrid interactors from Wormbase

- Closest homologs in other species and *C. elegans* (from BLAST searches)

- RNAi phenotype(s) (from large- and small-scale screens); make a note of the dsRNA delivery method (by checking the referral screen)

- Expression pattern

- Mutant strain available; phenotype of this strain

- GO annotations

- Significant findings from other species (such as disease connection, functional information or evolutionary importance) using literature searches

Much of this information can be found directly in Wormbase. Simply search by open reading frame number; the displayed results should contain most of the information requested above. You may have to visit some additional websites (i.e. PubMed, OrthoGo). After gathering all the information, discuss with your instructor which gene will be your RNAi target for the remainder of the course. If none of your sequenced clones are deemed of sufficient interest for further study, use the bioinformatics information available for Gene X to find a new related target suitable for RNAi. Alternatively, you could analyze the original Gene X.

Before you begin analyzing your sequence results, you may wish to practice accessing such information. Here we use the gene *nud-1* as an example to demonstrate the results. Listed below are the results to the above queries. Check the information that you obtain to see if you arrive at the same results.

- Open reading frame: F53A2.4

- Gene name: *nud-1*

- TopoMap Expression Mountain: 2 = RNA binding
  18 = Hermaphrodite enriched

- Yeast two-hybrid interactors: No hits

- Closest homologs: *nudC* (chicken)
  DnudC (fruit fly)
  RnudC (rat)
  nudC (human)

- RNAi phenotypes: Embryonic lethal (Emb)
  Everted vulva (Evl)
  Pronuclear migration defect (Pnm)
  Centrosome pair and pronuclear rotation defect
  Sterile progeny (Stp)
  Uncoordinated (Unc)
  Slow growth (Slo)
  Sterile (Ste)
  Exploded (Exp)

- Expression pattern: Amphid and phasmid sensory neurons
  Neuron ring and pre-anal ganglion
  Gonadal primordium
  Intestinal and hypodermal regions
  Early embryonic

- Mutant strain available: None

- GO annotations:
    Molecular function: Transcription factor activity
                        Unknown function
    Biological process: Embryonic development
                        Regulation of growth rate
                        Hermaphrodite genital morphogenesis
                        Regulation of transcription
                        Centrosomal and pronuclear rotation
                        Reproduction
                        Morphogenesis of epithelium
                        Unknown
                        Pronuclear migration
                        Locomotory behavior
    Cellular component: Unknown

- Significant findings: Role in hematopoietic cell growth
                        Role in leukemogenesis
                        Inhibits prostate tumor growth
                        Role in cytokinesis, mitotis, cell proliferation

6.4 Bioinformatics practice questions

The following questions are general tests of your bioinformatics skills. They range in difficulty and time investment but will all serve to provide you with further experience in using the databases previously discussed in this course.

(1) Obtain a copy of the following article: *Science*, 11 Dec. 1998, pp. 2012–2018.

(a) How many total genes is *C. elegans* predicted to have (based on this article)?

(b) At first examination, the *C. elegans* genome looks uniform with respect to GC content and gene density. However, differences do exist. Which region of the following chromosomal regions has the least distance between genes on average?

(A) The left arm of Chr. II

(B) The right arm of Chr. IV

(C) The right arm of Chr. III

(D) The central region of Chr. V

(c) Which of the following is correct with regard to protein matches to each other?

(A) *C. elegans* is more similar to *S. cerevisiae* than humans

(B) *C. elegans* is more similar to humans than *S. cerevisiae*

(C) *C. elegans* is more similar to *S. cerevisiae* than *E. coli*

(D) *C. elegans* is more similar to *E. coli* than humans

(2) (a) What is the name/number of the cosmid clone that the *dpy-3* gene is on?

(b) What are the last 10 amino acids at the end of the DPY-3 protein?

(3) In 1994, the lab of Dr Martin Chalfie developed the green fluorescent protein (GFP) as a marker for gene expression.

(a) In what journal was this published?

(b) Attach the complete DNA sequence for the plasmid encoding the version of GFP that was used in this paper.

(4) (a) What human protein has the highest amino acid sequence identity to the *C. elegans lin-10* gene product?

(b) What type of protein domain does LIN-10 share with this human protein?

(5) If a mouse researcher had the DNA sequence of a gene that she was interested in and she wanted to search for putative homologs in the zebrafish genome DNA database, which BLAST program would she use?

(A) blastx

(B) blastp

(C) tblastn

(D) blastn

(6) (a) Attach the results of a BLAST search for the identification of the 20 closest hits to the human SOD1 protein. You need provide only the sequence names and E-values (not the actual sequence matches). A typical threshold for E-values is e^{-5}.

(b) What human disease is linked to defects in this gene?

(7) Go to *NCBI Coffee Break* webpage. Open the *Archives*. Read the article entitled: '*PTEN and the tumor suppressor balancing act*'.

(a) What is the name of the closest *C. elegans* gene homolog of PTEN?

(b) Does this *C. elegans* homolog have a *higher* or *lower* probability of similarity to the human PTEN gene than the *Drosophlila* PTEN3 gene does?

(8) Using the *C. elegans* Expression Topographical Gene Map, list the open reading frame numbers of two genes that are co-expressed with F45H7.6 within a radius of 0.5.

(9) Using the *C. elegans* Topographical Gene Map, determine which of the following genes are potentially co-expressed with Y37A1B.13:

(A) C18F10.8
(B) F40E3.3
(C) F53A2.4
(D) Y37A1B.14

(10) Using Clustal 1.8, perform a multiple sequence alignment between the following proteins:

C. elegans MEC-2 protein
Erythrocyte Band 7 Integral Membrane Protein (stomatin)
Human stomatin-like protein 2 (SLP-2)
Predicted protein STO-3 from *C. elegans*

Attach the pile-up of sequences and highlight all identical amino acids shared across these proteins or use the Boxshade program to illustrate these residues.

7 *Generation of an RNAi vector*

After sequencing (Experiment 5) and identifying (Experiment 6) unknown cDNA products that may interact with your Gene X product, your next step of discovery involves a more detailed analysis of a few select targets for functional analysis by RNA interference (RNAi). RNAi is a technique used to study the phenotypic effects of knocking down the level of a gene transcript. It involves an overall reduction in the amount of transcript rather than the expression of a mutant protein (as is found in many genetic mutant strains). This chapter details traditional recombinant DNA methodology using restriction enzymes and ligation for generating a specific plasmid that will be used in RNAi. However, because molecular biology methods are constantly changing, this is no longer the only way in which to clone DNA segments into vectors. Recombinational cloning, which does not utilize restriction enzymes or ligations, is an alternative method to combine gene fragments with high efficiency. Protocols for cloning via recombinational methods can be found in Appendix I.

It is important to note before proceeding that the methods outlined in the previous chapters can only identify *putative* interacting partners. If you wished to further validate that two gene products are interacting, there would be additional experimental validations necessary (for example, *in vitro* pull-downs, co-immunoprecipitations, tandem affinity purification) that are outside the scope of this book.

7.1 Small-scale isolation of genomic DNA from *C. elegans*

In order to perform RNAi analysis, one must first clone a region of your gene of interest (hereafter 'Gene Y') into a specialized vector, known in the *C. elegans* field as

Integrated Genomics   Guy A. Caldwell, Shelli N. Williams, Kim A. Caldwell
© 2006 John Wiley & Sons, Ltd

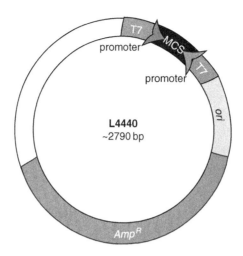

Figure 7.1   Diagram of vector L4440

plasmid 'L4440' (see Figure 7.1; detail in Figure AV.3). As will be discussed further in the introduction to Experiment 8, RNAi works by recognizing double-stranded RNA (dsRNA) specific to the target gene. This recognition triggers the enzymatic degradation of the dsRNA. When the RNAi machinery cuts the dsRNA, it will also degrade the single-stranded mRNA of the target gene naturally expressed within the organism and this results in reduction of the transcript and thereby reduction in the protein that arises from it. In order to perform RNAi, dsRNA must be generated against the specific target. The simplest way to create dsRNA is to let cells do the work for you. To synthesize dsRNA, you need only transcribe the DNA template on both strands at once. The bacteriophage T7 polymerase transcribes RNA from a DNA template. By placing your target DNA sequence between two bacteriophage T7 polymerase promoter sequences, you can employ the cellular machinery to work for you. The polymerase will recognize both promoters and transcribe in both directions, generating dsRNA. Before you can place your DNA template between the two promoters, however, you must first amplify the target sequence from the worm genome. This is done using a method called the polymerase chain reaction (PCR; see Section 7.2 for a review of this method). However, initially you must isolate genomic DNA from wildtype *C. elegans* to serve as the template for your PCR reaction.

Procedural overview

Isolating genomic DNA is a relatively simple procedure with *C. elegans*. Initially, you simply grow worms to high density on bacterially seeded agarose plates. It is important to use agarose, and not agar, when making these plates for several reasons. First and foremost, agar contains impurities that can affect the efficiency of the downstream PCR reaction. Additionally, agarose is much harder than agar and will prevent the loss of worms because they cannot burrow into agarose as they might into agar. To make agarose plates, simply substitute equal quantities of agarose for agar in the worm media recipe (Appendix II).

After growing several plates of worms to confluence (while ensuring that they do not run out of food), animals are harvested and washed several times to remove any impurities and bacteria. After washing, worms are digested for several hours in a detergent solution that also contains proteinase K, a non-specific protease that will degrade most of the proteins in the worms. The proteinase K is then heat inactivated and the mixture is stored frozen at $-20\,^{\circ}\mathrm{C}$ until use.

A crude preparation of *C. elegans* genomic DNA can be used several times before you will observe a decrease in the efficiency of the PCR. It is advisable to aliquot the crude preparation into several tubes so that the samples are not freeze–thawed more than twice.

Reagents (see Appendix II for recipes)

Agarose worm plates

Double-distilled water, sterile

Worm Lysis Buffer (complete with proteinase K, 19 mg/ml stock)

Equipment

Glass Pasteur pipets

Pipette bulbs

Glass centrifuge tubes, 15 ml

Table-top centrifuge

Microcentrifuge tubes, 1.5 ml, sterile

Microcentrifuge

$-80\,^{\circ}\mathrm{C}$ Freezer

$60\,^{\circ}\mathrm{C}$ Waterbath or heating block

$95\,^{\circ}\mathrm{C}$ Waterbath or heating block

$-20\,^{\circ}\mathrm{C}$ Freezer for storing crude DNA sample

Two to three days before lab

(1) Grow 3–4 plates of worms on agarose plates seeded with bacteria. Worms should be incubated (20–25 °C) until progeny appear and densely cover the plates (preferably just before food has been cleared; do *not* allow worms to grow to point of starvation).

Day of the lab

(2) Wash the worms off the plates with sterile double-distilled water by pipeting 2 ml onto one of the plates, then mixing up and down with a pipettor or glass Pasteur pipet and transferring them with this pipet to a 15-ml glass centrifuge tube. Repeat for each of the plates – combine all worms into the one 15-ml tube.

(3) Add sterile double-distilled water to 8 ml final volume. Spin in a table-top centrifuge for 2 min at 2000 rpm (no higher). Carefully remove most of the supernatant with a pipettor.

(4) Wash the worms with 5 ml of sterile double-distilled water and centrifuge again, as above.

(5) Remove most of the water, leaving approximately 500 μl of worms and double-distilled water. Transfer worms to a 1.5-ml microcentrifuge tube.

(6) Centrifuge the worms at 2000 rpm (no higher!). Carefully remove the excess water from the worm pellet (leaving worms as dry as possible without pipetting them off).

(7) Freeze worm pellet at −80 °C for at least 30 min.

(8) Add 40 μl of Worm Lysis Buffer to the pellet. Resuspend animals using a pipettor and incubate at 60 °C for 4 h.

(9) Heat inactivate the mixture at 95 °C for 20 min. This crude DNA preparation can be stored frozen at −20 °C until use. It is recommended that, before freezing, the crude DNA preparation is aliquoted into several tubes to prevent freeze – thaw-related degradation of your sample.

Bioinformatics

Exploring new databases: *Nucleic Acids Research* **database issue (http://nar. oupjournals.org/)** In this age of the information super-highway, bioinformatics resources are constantly growing. In some cases, it can be difficult to keep up with the latest updates to your favorite database, much less new databases. Different journals will often print updated lists or descriptions of the latest major change or newest addition to the bioinformatics realm. One such journal is *Nucleic Acids Research*, which publishes an annual issue discussing online databases for use by the scientific community.

Follow the address above to the link for the latest database issue by *Nucleic Acids Research*. Choose one database from each of the categories within this issue (i.e. Protein Sequence Motifs, Comparative Genomics, etc.). Write a brief review of each of these databases.

You may also find information on new bioinformatics databases in the 'Webwatch' section of *Biotechniques*, a free periodical for scientists, or in *Science* magazine's 'Netwatch'. Both of these journals frequently review new and updated bioinformatics tools.

7.2 PCR amplification of target gene sequence from *C. elegans* genomic DNA

Now that you have isolated *C. elegans* genomic DNA, you can proceed with a PCR reaction to amplify specifically your Gene Y genomic DNA sequence. In order for RNAi (see the introduction to Experiment 8) to work, you only need a relatively small region of DNA specific for your target to serve as a trigger for initiating RNAi in worms (i.e. the entire coding region is not necessary). A recommended target size is approximately 200–500 base pairs (bp) in length. RNAi targets the RNA sequence only. The presence of introns in your DNA template will not contribute to the overall RNAi efficiency, therefore primers should be designed that are specific to a region of DNA composed mostly of exons; it is not necessary to amplify *only* coding sequence. Amplifying only coding sequence would require a cDNA template rather than a crude genomic preparation (which is what you have isolated).

You will be adding a restriction enzyme site to your target specific primers in order to facilitate cloning. For a discussion on designing primers and choosing restriction enzymes to use in cloning, see the 'Instructor's notes' at the end of this section. Before proceeding, please review the brief PCR refresher provided in the 'Procedural overview' below.

Procedural overview

The polymerase chain reaction (PCR) is a method used to increase greatly the amount of a target DNA *in vitro*. Prior to 1985, when the PCR was invented, scientists used cells to replicate their DNA samples. The PCR is a useful tool not only for its rapidity (a typical reaction takes only a few hours) but also its specificity. To set up a PCR reaction, a DNA template is mixed with a supply of nucleotides, buffer, thermostable polymerase, and primers. The DNA template need not be from a highly purified source. In fact, DNA samples from a 10 000-year-old frozen mammoth and the 5000-year-old mummified Iceman have been used in successful PCR reactions! Primers are short synthetic DNA sequences that are complementary to the outer extremes of the target DNA sequence being amplified. A PCR is initiated when the double-stranded DNA template is separated (called denaturing) by heating the sample. The temperature is then lowered so that the primers can bind to the complementary sequences found on the single-stranded template (known as annealing). The thermostable polymerase, which is not damaged by the denaturation step, then extends the DNA strands by adding nucleotides to the ends of the primers (known as extension). The result is now two copies of the DNA target sequence. The PCR continues with a repetitive series of these denaturation, annealing, and extension steps, resulting in a logarithmic expansion in the amount of the target sequence. The last step is a final extension to ensure that all newly synthesized strands are fully extended before ending the PCR reaction.

Reagents

5′ Gene target specific primer (20 pmol/μl)

3′ Gene target specific primer (20 pmol/μl)

NOTE: Target genes will differ in this course. Be sure to make a note of YOUR specific target in your notebook as specified by the instructor.

Taq DNA polymerase (Fisher Scientific), 1 unit/μl stock concentration

Taq Buffer A

C. elegans genomic DNA (wildtype N2 variety)

dNTP mixture, 10 mM (these consist of dATP, dCTP, dGTP, and dTTP)
Double-distilled water, sterile

Equipment

0.2-ml PCR thin-walled tubes

Thermocycler, preferably enabled for gradients accepting 0.2-ml tubes, programmed as detailed below. (If a gradient thermocycler is not available, it is suggested to set the annealing temperature 3–5 °C lower than the predicted melting temperature (T_m) for the primer with the lower T_m; the T_m for a given primer is usually provided by the company synthesizing it)

Ice bucket

The PCR program

Step	Temp.	Time	Purpose of each step
1	94 °C	2:00 min	Preheat
2	94 °C	0:30 min	Denaturing
3	54–62 °C	1:00 min	Gradient Annealing*
4	72 °C	1:00 min	Extension
5	Go to step 2, 29×		Cycling
6	72 °C	7:00 min	Final Extension
7	End		

* Gradient predictions, with columns numbered from left to right:
1, 54.0 °C; **2**, 54.2 °C; **3**, 54.6 °C; **4**, 55.3 °C; **5**, 56.3 °C; **6**, 57.5 °C; **7**, 58.8 °C; **8**, 60.0 °C; **9**, 60.8 °C; **10**, 61.4 °C; **11**, 61.9 °C; **12**, 62.0 °C.

NOTE: The PCR is highly sensitive to contamination. Wear gloves throughout this procedure.

(1) Prepare 100 µl of both a 1:10 and a 1:100 dilution of the *C. elegans* wildtype genomic DNA that you isolated in Section 7.1. Place in an ice bucket.

(2) Preheat the lid of the thermocycler by initiating the program outlined above. When the lid temperature reaches 94 °C, pause the program.

(3) Label three sets of five 0.2-ml thin-walled PCR tubes: Set A (1–5), Set B (1–5), and Set C (1–5). Each set of tubes will correspond to a different set of DNA and will be used in a temperature-gradient PCR reaction.

(4) Prepare a "Master Mix" for PCR in a 1.5-ml microfuge tube by using the recipe in the table below, which has been calculated for both 1 and 15 PCR reactions. Because you will be performing 15 total reactions – five tubes for each source of template (1:10 dilution; 1:100 dilution; and undiluted) – this calculation is provided. Your Master Mix is equivalent to the *number of samples to be amplified + one extra* (to compensate for pipetting error).

	1 Reaction	15 Reactions (+1 extra)
10 × Taq polymerase buffer	5.0 μl	80 μl
10 mM dNTPs	1.0 μl	16 μl
20 pmol/μl 5 primer	1.0 μl	16 μl
20 pmol/μl 3 primer	1.0 μl	16 μl
Taq DNA polymerase	0.5 μl	8 μl
Sterile double-distilled water	40.5 μl	648 μl

(5) Mix your components completely and then aliquot 49 μl of Master Mix into each of 15 thin-walled PCR tubes.

(6) Add 1 μl of each dilution to a set of five 0.2-ml PCR tubes.

   Set A: 1 μl of a 1:10 dilution of your genomic DNA preparation
   Set B: 1 μl of a 1:100 dilution of your genomic DNA preparation
   Set C: 1 μl of your undiluted genomic DNA preparation

(7) Using a pipettor set at 40 μl, pipet several times to thoroughly mix the components. Change tips between samples.

(8) Place the reaction tubes in the thermocycler and close the lid. Resume the program by pressing the 'pause' button. This program takes ∼3 h. Samples should be removed and stored at 4 °C until the next class period. If your thermocycler does not have a heated lid, be sure to overlay your PCR mixture with PCR-grade mineral oil to prevent condensation from forming in the lid of the tube.

Bioinformatics

Restriction enzyme analysis using the Internet Turn to the 'Instructor's notes' at the end of this section. Find and read the part entitled 'Choosing restriction enzymes to add to your primers for cloning', which discusses how you would use online programs to help determine which restriction enzymes to use when ligating together two pieces of DNA. Being able to determine the restriction map of a sequence is a useful tool for any molecular biologist. For this experiment, your instructor performed this for you prior to ordering the primers you used for the PCR in the previous exercise.

   Review this part carefully; although some of the information applies specifically to what you needed for the previous exercise, it also provides basic information regarding mapping programs.

(1) Perform the sequence analysis suggested using your RNAi target and the mapping program of your choice. You will need to obtain an electronic copy of the target sequence. Use Wormbase if you do not already have this information.

(2) Using this same program, describe how many of the following restriction enzyme sites are present in a *C. elegans* corresponding to your Gene X interacting protein's cDNA

    (A) *Sac* I

    (B) *Bam*HI

    (C) *Hind* III

    (D) *Ava* I

Instructor's notes

Designing primers

In the era of modern bioinformatics wherein many different genomes have been completely sequenced, designing primers is a relatively simple process. To design primers for amplifying a target sequence, you must first access the sequence data for your target. You will find both spliced sequence (cDNA) and unspliced sequence (coding+introns) at www.wormbase.org by simply searching for your target gene (for example, *nud-1*) or open reading frame designation (F53A2.4).

When designing primers for the purpose of RNAi in *C. elegans*, you need only several hundred base pairs of specific sequence for your target (coding and intronic can both be included in this range but you *will* need at least 200–300 bp of coding sequence for successful gene knockdown). An example of this strategy is depicted in Figure 7.2 (you and your instructor should consult on designing your own primers). After finding the segment of genomic DNA that you wish to amplify (shown in box), record approximately 15–25 bp immediately upstream of your target region exactly as you see them displayed in Wormbase from left to right (shown in the light grey box in this figure). This segment will correspond to your 5′ primer. For your 3′ primer, copy 15–25 bp at the end of your target sequence (shown in the medium grey box) *but* then convert the sequence to the complementary sequence (shown in the dark grey box).

Some general guidelines to follow when designing primers are listed below.

(1) You should end your primer with a G or C because the triple hydrogen bond formed between the G or C in the primer and the complementary base in the template will ensure that your primer is tightly connected to the template at the site of polymerase extension.

1. Genomic region for gene targeting with flanking primer sequences

2. Bases upstream of targeted sequence corresponding to the 5' primer

5' g g a t t t c t a c a g t g t a t c g a c c c 3'
This primer is in the correct orientation for synthesis

3. Bases downstream of targeted sequence corresponding to the 3' primer

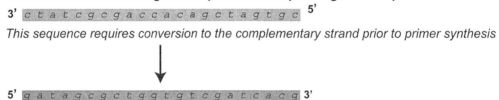

3' c t a t c g c g a c c a c a g c t a g t g c 5'
This sequence requires conversion to the complementary strand prior to primer synthesis

5' g a t a g c g c t g g t g t c g a t c a c g 3'
This primer is now in the correct orientation for synthesis

Figure 7.2 Example of primer design for PCR amplification of RNAi target sequence

(2) Design primers so that they are approximately 50% G/C. Overly G/C-rich sequences have a much higher melting temperature and may decrease the efficiency of primer annealing.

(3) When designing primers to be used in the same PCR, be sure that their G/C content is very similar. Otherwise, their melting temperatures will be too different for successful amplification.

(4) It is best to avoid long stretches of either As and Ts or Gs and Cs. Again, these regions can affect the specificity and efficiency with which the primer binds to the template.

(5) Try to avoid internal hairpin sequences. These are sequences where a stretch of several bases is complementary to a stretch of several bases at the other end of the primer. These two regions may bind to each other (producing a 'hairpin' shape) instead of to your template.

Alternatively, many programs exist online that will design primers for you. A few sites are listed below but a simple search engine query of 'primer design programs' will yield many hits so that you may choose your own program. Please follow the input directions of the specific program you are using.

www.invitrogen.com/oligos
frodo.wi.mit.edu/cgi-bin/primer3/primer3_www.cgi
seq.yeastgenome.org/cgi-bin/web-primer
bibiserv.techfak.uni-bielefeld.de/genefisher/

After you have designed your primers, complete with sequences necessary for cloning (see under next heading), it is necessary to obtain the primer. Synthetic oligonucleotides (such as primers) are available from a number of sources. Other members of your institution can likely recommend a company for purchasing primers. A few commercial sites are listed below for ordering primers; again, a simple Internet search will yield numerous additional options for ordering primers. It is recommended that you order primers as soon as possible following sequencing of your yeast two-hybrid clones to minimize delay in the course. Companies that produce primers should be able to tell you their average turnaround time on orders (typically 3–4 days).

www.invitrogen.com
www.mwgbiotech.com/html/order/login.shtml
www.openbiosystems.com
www.operon.com

When your primers arrive, they will likely be in lyophilized form (completely dried to a powder). You should reconstitute your primers in TE buffer, pH 8.0. Primers are generally maintained in a stock solution of 100 pmol/μl; the information provided with the primers should tell you what volume to use for resuspension. This stock is then diluted using sterile, or preferably molecular grade, water to the working concentration of 20 pmol/μl.

Choosing restriction enzymes to add to your primers for cloning

Choosing restriction enzyme sites for cloning is slightly more complicated than designing primers. In this case, you must use a computer program. First, you will need a copy of the target sequence that you wish to clone into a vector and a restriction enzyme map of your vector (in our case, plasmid L4440). As detailed in the vector map (see Figure 7.1), most plasmids contain a region known as the Multiple Cloning Site (MCS). This is a specific region of the vector that contains many restriction enzyme sites that are typically unique within the plasmid. After you have gathered the RNAi target sequence information from Wormbase, simply copy your target DNA sequence into a restriction enzyme mapping program. Several are listed below, but a simple search engine query will find alternative options. Choose the one you like best and follow its directions.

www.restrictionmapper.org/
www.firstmarket.com/cutter/cut2.html
tools.neb.com/NEBcutter2/index.php
www.mekentosj.com/enzymex/

When you enter your sequence into the mapping program, you will likely have several options for type of output. Although it is helpful to know a given sequence's entire restriction map, for cloning purposes you need only determine which enzymes do *not* cut the sequence. When you have this information, compare the enzymes that do not cut your selected DNA fragment with the available sites in the MCS of the vector L4440. It is highly desirable to choose two different restriction enzymes so that your digestion will yield directional cloning and avoid self-ligation of your vector. In other words, if you know which enzyme site you placed on the 5′ end, you will know in which direction your DNA was ligated into the vector. After you have chosen the restriction enzymes that you will use, go to the restriction enzyme catalog or website for your source of enzymes (i.e. New England Biolabs). Verify that your chosen enzymes are compatible in terms of the buffers used (look in a double digestion chart) and the working temperature, and that they do not overlap within the vector. If the two enzymes exhibit sufficient levels of digestion in the same buffer (50 percent or greater) and work at the same temperature, those enzymes represent good choices for incorporation of their recognition sequences into your primers.

Now that you have determined which enzymes to use, you will need to update your primer sequences. Under each restriction enzyme listed in your catalog you will find the recognition sequence for that enzyme. For the enzyme site that you have chosen to place at the 5′ end of your sequence according to the orientation in the MCS, you will need to add that recognition sequence directly to the 5′ end of your 5′ primer. In other words, add the recognition sequence to the 5′ end of your primer (left to right, directly in front). It would then read: 5′-recognition sequence-target sequence-3′. For the enzyme you will add to the primer used to amplify the 3′ end of your target sequence, you will need to add the recognition sequence in *reverse complementary* order at the 5′ end of your 3′ primer. For example, you have chosen 400 bp of target gene sequence. Looking at the Wormbase sequence, your region reads atgctacccgtaatacgggtatgcatgattacgtat . . . (additional 330 bases). . . gttaacatgtcatgtcatgatcatgtacctagtat from left to right on the page. You want to add the recognition sequence for *EcoRI* (recognition sequence GAATTC) to the 5′ end of your target sequence and for *XbaI* (recognition sequence TCTAGA) to the 3′ end of your target, therefore you would design the following primers:

5′ primer: 5′-GAATTCatgctacccgtaatacgggtatgcatg-3′

3′ primer: 5′-AGATCTatactaggtacatgatcatgacatg-3′

Note that the 3′ primer has the *XbaI* site and the target sequence in reverse complementation to what you see. Again, remember that you are amplifying double-stranded

DNA and that the 3′ primer and restriction enzyme site corresponds to the 5′ end of the 'bottom' strand. If you have difficulties, many primer design programs will also allow you to add restriction sites chosen from a list.

A final note on adding restriction enzyme recognition sequences to primers: restriction enzymes are proteins, which are three-dimensional structures that attach to specific sequences of DNA. Ordinarily, restriction enzymes cut DNA sequences that are contained within larger stretches of DNA (i.e. the whole genome or chromosome). The flanking bases around the recognition sequence add stability to the enzyme and its ability to bind the DNA. Many enzymes work better if the recognition sequence added to the end of the primer also contains flanking bases. Obviously, your DNA target will add bases to the 3′ end of the recognition sequence. To determine if your enzyme requires extra bases on the 5′ end of the recognition sequence, consult the catalog of a commercial manufacturer such as New England Biolabs. You should be able to find a section in the appendices that tells you the efficiency of cleavage close to the end of the DNA fragments. This information will tell you how many, if any, bases you need to add before the recognition sequence. The identity of these bases does not matter; they merely serve as a platform for enzyme activity.

If none of the restriction enzymes listed within the MCS of L4440 will work for your chosen sequence (i.e. if they all cut that region of DNA), you will need to choose a different region within your gene to target for RNAi and begin again with primer design.

7.3 Preparations for cloning to generate RNAi vector

After you have performed a PCR reaction, it is necessary to determine if the reaction was successful. This is done by examining a portion of the PCR sample using agarose gel electrophoresis (see Section 7.3.1). Once you have obtained the correct size amplification product, you are almost ready to proceed with the first step in cloning – the restriction enzyme digest (see Section 7.3.3). Before the digest however, you will clean up your PCR reaction to remove unincorporated nucleotides, primers, and contaminants using a spin column (see Section 7.3.2).

Reagents (see Appendix II for recipes)

Agarose (powder)

1 × TAE buffer

Ethidium bromide, 0.625 mg/ml (CLP dropper bottle 5450) (*Careful: it is a mutagen!*)

DNA molecular weight marker, 1 kb ladder (New England Biolabs) containing 1X final concentration gel loading dye

6X Gel loading dye: this should be used as 1/6 of the final volume of any DNA solution to be electrophoresed (i.e. add 1 μl of 6X dye to 5 μl of DNA)

QIAquick PCR purification kit (Qiagen)

Restriction enzymes 1 and 2, 10 U/μl

100X Bovine serum albumen, if required by either enzyme

Restriction enzyme buffer, appropriate for use by both of your enzymes (buffers appropriate for use with two enzymes can be found in the New England Biolabs catalog or online under the 'Double Digestion' reference appendix or the NEBuffer Activity Chart)

Equipment

Glass flask with boiling cap

Microwave

Casting trays or laboratory tape for pouring gels for electrophoresis

Parafilm

Gel comb with appropriate number of wells

Gel trays, 9–10 cm size

Gel electrophoresis apparatus

UV illumination source

Gel documentation system (such as a camera attached to UV illumination)

Microcentrifuge tubes, 1.5 ml, sterile

Microcentrifuge tubes, 0.5 ml, sterile

Microcentrifuge

Water bath or heating block, set at appropriate temperature for restriction enzyme digest (typically 37 °C)

7.3.1 Agarose gel electrophoresis

Procedural overview

Agarose gel electrophoresis separates DNA based on size during the application of an electrical current. Following reconstitution in buffer, the agarose is melted in a microwave and then poured into a mold (a gel tray) where it solidifies into a porous substrate in ~20–30 min. Gels should be poured into trays whose ends have been carefully taped completely closed or that have been placed in casting trays. Once solidified, the agarose gel is placed in an electrophoresis chamber containing running buffer, which will ensure the even distribution of the electrical current when it is applied. Be sure to remove the tape if used to seal the ends during casting. To prepare your PCR sample for electrophoresis, a portion of the PCR sample is mixed with a loading dye containing Ficoll, which weighs down the DNA, ensuring that it does not float out of the wells. The dye will also provide a visual cue as to how far the DNA has migrated through the gel. When the dye and DNA have been added to the gel, the electrophoresis chamber is closed and an electrical current is applied.

When loading an agarose gel, it is very important to be sure that your gel and DNA are positioned correctly. The DNA should be loaded near the negative electrode of the chamber. Because like charges repel and opposites attract, the negatively charged DNA will migrate through the gel toward the positive electrode due to the high phosphate content of the DNA backbone. As the DNA moves through the porous agarose, smaller pieces will migrate more rapidly than larger pieces, therefore when you examine the gel the smaller fragments will be closer to the positive electrode (the 'bottom' of the gel) and the larger pieces will be nearer to the negative electrode (the 'top') (see Figure 7.3).

It is also important to include a DNA molecular weight marker in each gel that you run in a lane adjacent to your samples. This serves as a ruler to determine the size of the DNA band you are examining. It is important to verify the size of your anticipated PCR product for your target gene to ensure that you have obtained the correct amplification product.

Lastly, agarose gels can be prepared with ethidium bromide (EtBr) stain incorporated into the agarose or the gel can be submerged in a solution containing EtBr after the DNA sample has been run through the gel. *Handle EtBr very carefully because it is a mutagen.* In this procedure the EtBr is added directly to the gel: *wear gloves when handling the gel*!! Following electrophoresis, the results of the experiment are documented by photographing the UV-illuminated gel.

(1) Weigh out 0.8 g of agarose and place it into a 500-ml flask. Add 100 ml of 1X TAE buffer and microwave for 2–3 min, bringing the solution to the boil. Swirl the

DNA fragments of different sizes are loaded into the wells of
the gel

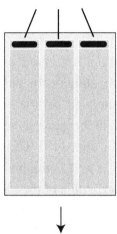

An electrical current is passed through the gel and DNA
fragments migrate toward the positive electrode

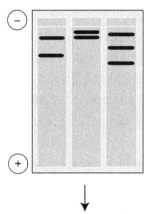

The fragments are separated by size with the smaller ones
migrating faster than the larger ones. These DNA bands are
visualized with a dye that specifically binds to nucleic acids

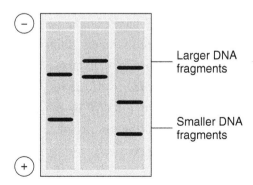

Figure 7.3 Separation of DNA fragments in an agarose gel during electrophoresis

solution and microwave again, if necessary until all 'refractile fish' are dissolved. *Wearing gloves*, add two drops of EtBr (a mutagen) into the flask and swirl the solution to diffuse the EtBr.

(2) Allow agarose to cool on the benchtop for 10–15 min and pour into a gel tray that has either been taped on the sides or placed in a casting tray. Insert the gel comb at the notches closest to one end of the gel.

(3) Allow the gel to solidify for ∼30 min.

(4) Place the gel tray into an agarose gel electrophoresis apparatus. Orient the gel so that the wells are adjacent to the negative (black) electrode. Fill the gel chamber with $1 \times$ TAE buffer until it covers the gel by ∼1 cm. Carefully, and evenly, pull the comb out of the gel using both hands.

(5) Prepare PCR samples for loading by pipeting a 1-μl spot of $6 \times$ gel loading dye for each sample to be loaded onto a piece of Parafilm spread on your benchtop. *While noting the orientation of your samples*, pipet 5 μl of each PCR sample into each spot of dye. Be sure to change pipet tips between samples and do not contaminate the PCR sample with gel dye. For each sample, mix up and down with the pipettor and directly load it all into a well of the agarose gel. Load 0.5 μg of 1 kb molecular weight ladder into an empty lane.

(6) Attach the electrode cover to the gel electrophoresis unit such that the positive (red) electrode is at the opposite end of the gel from the sample wells. DNA is negatively charged and moves toward the positive electrode. Plug the electrodes into the power supply and set it for 100 volts. Start the electrical current. The sizes of available electrophoresis units vary, but a run will take approximately 45 min to 1 h for the DNA to separate in a 9-cm gel. Using the DNA dye recipe in this procedure, you should see two bands of migrating color: bromophenol blue migrates to a position approximately euqivalent to 300 bp of dsDNA, whereas xylene cyanol FF migrates at 4000 bp. Be sure to stop the current before the bottom line of gel dye migrates out of the gel.

(7) *Wearing gloves*, remove the gel tray from the apparatus for documentation. You may want to carry the gel in another container (i.e. baking tray or plastic dish) because sometimes they will slip out of the gel tray. A broken gel yields no data. Follow instructions for use of the specific photography/gel documentation set-up available to your lab and print a copy of your gel picture. Record data and interpret the success of your PCRs. A successful PCR should give you a single band of the appropriate size based on the primer design. Your instructor should inform you of the expected size of your amplified DNA fragment. Dispose of gel in an appropriate hazardous container.

(8) If your PCR reaction was unsuccessful, you will need to repeat the exercises in Sections 7.2 and 7.3.1 before continuing. Consider altering the template concentration and/or annealing temperature to generate an appropriately sized PCR product. See 'Instructor's notes' at the end of Section 7.3 for PCR troubleshooting or buying RNAi clones.

7.3.2 Removal of dNTPs from PCR reaction

Once you have verified that you have successfully amplified a PCR product of the correct size, it is necessary to remove all contaminants from the solution containing this product. These include remaining nucleotides, buffer, polymerase, and genomic DNA. This is necessary to ensure that contaminants do not interfere with the subsequent restriction enzyme digest you will perform on the amplified product. Clean-up is performed using a silica resin spin-column very similar to those used in the mini-prep procedure (see exercise in Section 5.3).

Procedural overview

Begin by adding Buffer PB to the PCR product. This buffer, which contains a high concentration of chaotropic salts, efficiently binds the PCR products to the filter when the solution passes through the column during centrifugation. The filter is then rinsed with Buffer PE, an ethanol-based solution that serves to remove contaminants from the PCR. The column is centrifuged an additional time without buffer in order to dry the sample before elution is performed in Buffer EB, a low-salt, high-pH solution optimized for solubilizing DNA. This solution serves to elute the DNA from the filter, resulting in a purified sample suitable for many downstream applications, including restriction enzyme digestion.

Qiagen QIAquick PCR purification kits are capable of purifying double-stranded PCR products ranging from 100 bp to 10 kb in size. Should you ever need to purify larger products, it will be necessary to acquire a different kit or utilize alternative methods such as phenol – chloroform extraction followed by ethanol precipitation.

NOTE: The instructor may ask you to combine two or more successful PCR samples prior to initiating this purification procedure. Ask prior to proceeding. This is strongly advised if the intensity, of the PCR products of the successful amplifications is not robust. Typically, combining 4–5 PCR reactions is acceptable to concentrate a given product prior to subsequent use. Be aware, however, that the maximum loading volume of the column is 800 µl. Again, note that this procedure recommends the use of a Qiagen spin column system. Follow the manufacturer's instructions if using a different type of clean-up kit.

(1) Transfer your PCR reaction to a 1.5-ml microcentrifuge tube and add five volumes of buffer PB to the reaction. Vortex to mix. For example, for a standard 50-µl PCR you should add 250 µl buffer PB.

(2) Place a QIAquick spin column in the provided 2-ml collection tube and apply the entire sample from Step 1 to the center of the filter.

(3) Centrifuge for 1 min at top speed in a microcentrifuge.

(4) Drain the flow-through fraction from the collection tube by pouring it into a waste container. Retain the collection tube and re-use for the next steps.

(5) Wash the QIAquick spin column by adding 750 µl of Buffer PE and centrifuge for 1 min at top speed in the microcentrifuge.

(6) Drain the Buffer PE flow-through from the collection tube by pouring it into a waste container. Again, retain the collection tube.

(7) To dry the column, centrifuge the QIAquick spin column for an additional minute to remove residual Buffer PE. Discard the collection tube.

(8) Place the QIAquick spin column into a clean 1.5-ml microcentrifuge tube.

(9) Elute DNA by adding 30 µl of Buffer EB directly onto the center of the filter. Allow the elution buffer to permeate the resin for 1 min.

(10) Collect your DNA sample by centrifuging the column for 1 min.

(11) Quantitate each sample using a DNA spectrophotometer. Refer to the 'Instructor's notes' in Section 5.4 for information on quantitating DNA.

(12) Store samples at 4 °C until use in restriction enzyme digestion.

7.3.3 Restriction enzyme digestion of PCR product and *C. elegans* RNAi vector

Now that you have purified your PCR product, it is time to proceed with a restriction enzyme digestion to facilitate cloning of your PCR product into plasmid L4440 used for RNAi. Restriction enzyme digestion has been commonly used for recombinant DNA cloning since the advent of this technology in the 1970s. Many restriction enzymes are commercially available for use; each works by recognizing a very specific DNA sequence and hydrolyzing the bonds between nucleotides at that site. When the DNA is cut, a specific 'overhang' often remains. If two pieces of DNA are cut by the same restriction enzyme, their overhang patterns will match (see Figure 7.4). This similarity enables DNA molecules to be recombined. A DNA repair ligase is then added to the reaction, enabling the complementary overhangs to be sealed together, generating a single piece of DNA from what was once two pieces. These overhangs are also known as 'sticky ends', meaning that the ends will stick together following the ligation reaction. Ligation is discussed further in the next section. By using two different restriction enzymes, one for each end of a PCR product or DNA fragment, researchers may control

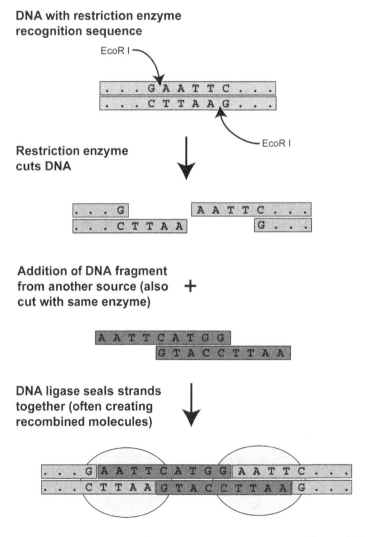

Figure 7.4 Schematic representation of cloning a fragment using a single restriction enzyme

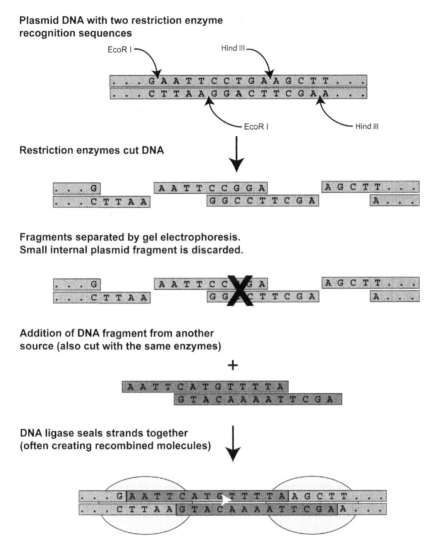

Figure 7.5 Schematic representation of cloning a fragment using two different restriction enzymes

the direction in which the two pieces of DNA are ligated together (see Figure 7.5). This enables orientation of a gene within a vector, which can be important for some studies, e.g. when a start codon (ATG) must be in a specific orientation within a vector to facilitate proper gene expression from an upstream (5′) promoter element.

In order to perform a restriction enzyme digestion, you need only mix together the proper components and incubate the reaction appropriately. When restriction enzymes are purchased, buffers designed for optimal DNA cleavage are provided by the manufacturer. Prior to performing a digest using two enzymes, verify that each will work well in a single buffer. Many enzyme suppliers provide charts illustrating which buffer to use for different combinations of commonly used enzymes. It is desirable to investigate the capability of the specific restriction enzymes utilized in recombinant DNA cloning prior to incorporating their recognition sequence into the PCR primers used to amplify a given target. Please refer to the 'Instructor's notes' in Section 7.2 for further details on choosing restriction enzymes and recognition site incorporation.

When the DNA, buffer, and enzymes have been mixed together, simply incubate the tubes at the appropriate temperature overnight. Most restriction enzymes have an optimal reaction temperature of 37 °C; however, occasionally some enzymes have been determined to function best at alternative temperatures. It is wise to verify this in the manufacturer's catalog prior to setting up the reaction and to confirm that your two enzymes work at the same temperature. Be sure that you set up one digestion reaction containing the PCR product and a second containing the target vector, in this case plasmid L4440. Digests should not be allowed to proceed too long because the enzymes may lose their specificity over time and begin digesting DNA indiscriminately (this is referred to as star activity). If your lab does not meet every day, it is recommended that the digests be started 12–18 h before the exercise in Section 7.4, either by the students or by the instructor. Carefully record in your notebook who set up the reaction and when it was started.

(1) Label two 0.5-ml microcentrifuge tubes. Pipet the following into each tube (in this order):

	For PCR fragment	For vector
DNA*	20.0 μl	10.0 μl
Sterile water	21.5 μl	31.5 μl
10 × Enzyme buffer	5.0 μl	5.0 μl
100 × BSA	0.5 μl	0.5 μl
Enzyme 1**	1.5 μl	1.5 μl
Enzyme 2**	1.5 μl	1.5 μl
Total	50.0 μl	50.0 μl

* Although exact quantities of PCR fragment and vector DNA could be specified (and may be preferred by your instructor), a range of 1–5 μg of starting material for each DNA sample generally yields enough product for successful ligation.
**When handling enzymes be careful to keep the enzymes on ice and return them immediately to the freezer after use!! Enzymes are stored in glycerol and are thus viscous solutions that one must pipet carefully.

Mix the tubes well by flicking the base with a fingertip *(do not vortex)*.

(2) Incubate at 37 °C overnight. (Most, but not all, enzymes function at 37 °C. Check the information that came with the enzyme carefully for the appropriate activity temperature.)

(3) Heat inactivate enzymes at 80°C for 20 min in a heat-block or heated water-bath. *Note*: Not all restriction enzymes are deactivated by heat denaturation – see manufacturer's information.

(4) If reactions must be left longer than 18 h, they should be placed at −20°C until you are ready to proceed with the exercise in Section 7.4.

Instructors' notes

PCR troubleshooting

The PCR is perhaps the most widely used technique in modern biological research. Its simplicity and sensitivity make it an invaluable tool for all types of research, from cloning novel genes and forensic analysis to examining evolutionary relationships across diverse species lineages. However, as with all techniques, it does not always work the first time. Several factors can influence the efficiency of a PCR, including the size of the desired amplification target, the purity and quantity of DNA template used, and minute changes in salt concentration within the solution, to name only a few. As mentioned in our procedure, the availability of gradient thermocyclers enhances the chance of amplifying a given target sequence. Thus, we *highly* recommend the use of a thermocycler capable of an annealing gradient. If your PCR does not work the first time, consider one of the following changes in subsequent attempts.

No visible product

(1) If possible, use a DNA template that you know has worked previously for the PCR. The crude preparation of genomic DNA from *C. elegans* used in this text can be variable with respect to template concentration and impurities (i.e. nucleases, excess proteinase K, etc.). If possible, a fresh preparation of worm genomic DNA is optimal for repeat PCR attempts.

(2) Decrease the annealing temperature for the primers so that they have a better chance of binding to the template.

(3) Alter the salt concentration in the buffer used. Various suppliers of *Taq* DNA polymerase offer different buffers for use with their enzymes.

(4) Choose a different target region within your gene; redesign primers appropriately.

Non-specific products visible

(1) If you see very short bands at the bottom of your gel (well below the last band of the ladder) these are most likely 'primer dimers', which form when the primers anneal to each other and not the DNA template. To reduce primer dimers, lower the amount of each primer used in the reaction. For example, try 10 pmol/μl. However, if you do have the correct size product *and* primer dimers, there is no need to repeat the PCR. Primer dimers are frequently purified away from larger DNA products during spin column clean-up.

(2) If you see multiple bands on your gel, including one of the correct size, your primers are binding non-specifically. Increase the annealing temperature by 2–5 °C to increase the specificity of your reaction.

(3) As above, altering the salt concentration and primer sequences will have a dramatic effect on your success.

Staining agarose gels with EtBr

Some researchers prefer to stain their agarose gels with EtBr after running the gel rather than including the stain in the solution itself while this is more time consuming it typically yields higher resolution of fluorescence for visualizing bands in gels. If you would prefer to do this:

(1) Prepare a working solution of EtBr by adding 50 μl of 10 mg/ml EtBr to 200 ml of 1X TAE in a plastic container.

(2) Submerge agarose gel in this solution for 30 min.

(3) Wash the gel in double-distilled water for 1 min to remove EtBr from the surface. Pour off water/EtBr into a waste container.

(4) Destain gel for 10 min in double distilled water.

Your gel is now ready for photodocumentation as detailed in the protocol above. The original EtBr solution can be stored in the dark and reused.

Buying RNAi clones

If time does not allow you to clone your own RNAi target into the L4440 vector or if your attempts are unsuccessful, clones are available for purchase. Individual *C. elegans* RNAi clones may be ordered from the following companies and are provided in bacterial hosts. Please visit their websites for prices and ordering information. Clones ordered from these companies are ready for use in an RNAi feeding experiment. Follow the instructions received with the clones for propagation and proceed immediately to the exercise in Section 8.4 for use in this laboratory scheme. It should be noted that at the current time not all genes are available for purchase.

- Geneservice
  http://www.geneservice.com/co.uk/

- Open Biosystems
  http://www.openbiosystems.com/rnai/

7.4 Gel purification of DNA and ligation of vector and PCR-amplified DNA

Following the restriction enzyme digest, you will have generated DNA with compatible sticky ends. However, before you can proceed with a ligation reaction (discussed below), you need to remove the enzymes and buffers used during the restriction digest. At this time, you no longer want these enzymes to digest the ends of your DNA because it is now essential to combine and insert your PCR fragment within the L4440 vector.

In order to remove the buffers and enzymes from your DNA samples, you will perform preparative agarose gel electrophoresis. Logistically this does not differ from the gel electrophoresis performed in the previous exercise. Your samples will be added to the gel and separated by size during electrophoresis. However, in this case you will add the *entire* digestion reaction to the gel. In this way, you are able to separate physically the DNA from enzymes and other contaminants within the reaction mixture, thus purifying and preparing the DNA for ligation.

Reagents

Agarose (powder)

1X TAE buffer

Ethidium bromide, 0.625 mg/ml (CLP dropper bottle 5450) (*Careful: it is a Mutagen!*)

DNA molecular weight marker, 1 kb ladder (New England Biolabs) containing 1X final concentration gel loading dye

6X Gel loading dye

Double-distilled water, sterile (preferably Molecular Biology Grade water; Qbiogene catalog no. 2450-204)

Qiagen MinElute gel extraction kit

Isopropanol

10X T4 DNA ligase buffer (New England Biolabs)

T4 DNA ligase (New England Biolabs)

Equipment

Glass flask with boiling cap

Microwave

Casting trays or laboratory tape for pouring gels for electrophoresis

Parafilm

Gel comb with appropriate number of wells

Gel trays, 9–10 cm size

Gel electrophoresis apparatus

UV illumination source

Razor blades, new

Microcentrifuge tubes, 1.5 ml sterile

50 °C Waterbath

Vortex

Ice bucket

Microcentrifuge tubes, 0.5 ml sterile

16°C Waterbath

7.4.1 Preparative agarose gel electrophoresis

(1) *Wearing gloves*, prepare a 0.8 percent preparative agarose gel in 1X TAE with EtBr (1 drop) by dissolving 0.5 g of agarose in 60 ml of 1X TAE buffer using a microwave and then pour into a gel mold (see Section 7.3.1). You will need to

tape *two or more* wells of the gel comb together for each sample to be gel purified in order to have wells large enough to hold your entire samples (60 µl following addition of dye). When taping the wells together, use clear tape to generate smooth wells rather than the colored laboratory tape. Pour the agarose into the mold and allow to solidify.

(2) Place solidified gel into the electrophoresis apparatus and cover with 1X TAE buffer.

(3) Add 10 µl of 6X gel loading dye directly into the tubes containing the digestion samples, mix well, and load the entire 60-µl volume into the wells formed from the taped comb slots. Load 6 µl of molecular weight marker (1 kb ladder) into another well.

(4) Electrophorese gel at 80–100 V for approximately 45 min to 1 h.

(5) Proceed *immediately* to the next exercise.

7.4.2 Gel purification of DNA from agarose gel

At this time it is necessary to purify your DNA from the agarose gel matrix to separate it from the components of the restriction digestion reaction and to enable isolation of your insert DNA and cleaved plasmid. This is performed using another filter-column system, the Qiagen MinElute gel extraction kit. This kit uses the same principles as the QIAquick PCR purification kit you used in the exercise of Section 7.3. The DNA initially binds to the silica filter under high salt conditions and contaminants are removed by a series of buffer washes. DNA is ultimately eluted from the silica filter by an elution buffer of basic pH and low salt concentration.

Procedural overview

You begin this procedure by physically cutting your DNA samples from the agarose gel using a razor blade. This is performed on a UV transilluminator. Ideally, attempt to cut as close to the band as possible to minimize excess agarose, which reduces the efficiency of DNA extraction. Be very careful during this step to stay behind a UV shield, either attached to the transilluminator or in the form of a face guard. Be *very* sure to wear gloves and UV-resistant goggles to protect your eyes and skin. This step should also be performed as quickly as possible to prevent UV damage to your DNA samples.

Following gel excision of your DNA samples, you will add the appropriate amount of Buffer QG. This buffer contains an ionic liquid that both solubilizes the agarose when heated and provides the appropriate environment for binding the DNA to the filter column. After the agarose is completely melted, the sample is added to a filter column. The DNA binds the filter as the contaminants (buffers, agarose, EtBr) wash through during centrifugation.

The filter is then washed using Buffer PE, just as it was during the PCR purification procedure, and the DNA is eluted with Buffer EB and collected by centrifugation. Elution in a small volume ensures that your cleaned DNA is more concentrated, which is an important consideration because the subsequent ligation reaction is performed in a small reaction volume to facilitate recombination.

NOTE: This procedure should be performed for both the vector and PCR fragment samples from your restriction enzyme digestion reactions in separate tubes (exercise in Section 7.3.3)

(1) Record the weight of two empty 1.5-ml microcentrifuge tubes by writing the weight on the tubes. Label each tube with the name of the fragment that will be isolated.

(2) Place gel on transilluminator and, *wearing gloves and protective glasses*, cut out the band(s) of interest using a clean razor blade while viewing with UV light. It is desirable to do this *quickly* ($< 30\,s$) to avoid nicking the DNA in the presence of UV light. Immediately turn off the UV light when the bands have been excised.

(3) Place the DNA-containing agarose into a preweighed tube. Weigh the tube again, record the weight, and subtract it from the original tube weight. If the agarose weighs more than 0.4 g it should be divided into two tubes (and the weight of each agarose piece recalculated). In this case, both tubes should undergo Steps 4 and 5.

(4) Add *three volumes* of Buffer QG. For example, 0.6 ml of QG buffer should be added to 0.2 g of DNA-agarose. Vortex the tube and place in a waterbath at $50\,°C$ to dissolve agarose. Incubate for 10 min, occasionally vortexing to facilitate the dissolution of the agarose. The gel fragment must dissolve *completely* before you proceed. Heat longer if necessary.

(5) If your DNA fragment is between 500 bp and 4 kb, proceed immediately to Step 6. If your DNA band is <500 bp or >4 kb, add one gel volume of isopropanol to the sample and mix. For example, use $200\,\mu l$ for a 0.2-g agarose sample.

(6) Add the sample to a MinElute spin column and centrifuge for 1 min at top speed in a microcentrifuge. Discard the flow-through, but keep the collection tube and column. The maximum volume that the column can contain is $800\,\mu l$. If your volume is greater than $800\,\mu l$, load and spin again until the entire volume has been added

to a single column. Discard the flow-through between spins. If you split your sample in Step 3, add all of the samples to the same column by consecutively adding 800-μl amounts and centrifuging until all the liquid has passed through a single filter.

(7) Add 750 μl of Buffer PE to the column. Centrifuge for 1 min at top speed.

(8) Discard the flow-through, but keep the collection tube and column.

(9) To dry the filter, centrifuge the column for an additional 1–2 min at top speed.

(10) Place the column into a 1.5-ml centrifuge tube. Add 10 μl of Buffer EB directly to the center of the filter. Watch carefully to ensure that the solution does not cling to the side of the column. Incubate at room temperature for 1 min.

(11) Centrifuge the column for 1 min at top speed. The solution contained within each microcentrifuge tube represents your purified DNA. Discard the filter column and proceed with the next exercise.

7.4.3 Ligation of vector and PCR-amplified DNA

Now that you have successfully purified your DNA samples, you are ready to proceed with the ligation reaction. In cells, DNA is often damaged during the course of everyday life and upon exposure to harmful environmental conditions (e.g. UV light or oxidation reactions). This often results in nicks or breaks in DNA, therefore a system has evolved that functions to repair and seal these breaks in DNA. Ligases catalyze the formation of a phosphodiester bond between the 5′ phosphate and 3′ hydroxyl ends of nucleotides. Molecular biologists have found that the bacteriophage T4 ligase is very efficient at this process and therefore employ this purified enzyme for *in vitro* reactions. Complementary segments (sticky ends) are recognized as fitting like puzzle pieces and are 'glued' back together by DNA ligase. This enzyme uses ATP as a cofactor during the reaction. Be very sure when setting up your ligation reaction that no white flakes are visible in the buffer solution, because this is undissolved ATP that would therefore not be available for use by the ligase. To ensure complete dissolution of the ATP, thaw the buffer at least 20 min before setting up the reaction and mix well before dispensing.

In addition to setting up a ligation reaction between your PCR fragment and plasmid vector, you will also perform a control reaction containing only plasmid DNA. This tests the efficiency of your restriction enzyme digestion. If you used two different restriction enzymes in your digestion, the vector should not be able to ligate closed upon itself. This will be apparent following bacterial transformation in the exercise of Section 7.5. The plasmid-only control reaction should yield few or no transformants because bacteria are not efficiently transformed by linear pieces of DNA.

(1) The following 15-μl reactions should be set up in two 0.5-ml tubes on ice (add components in the order listed):

	Ligation reaction	Plasmid-only control
Sterile double-distilled water (preferably Molecular Biology Grade)	2.5 μl	7.5 μl
10 × T4 DNA ligase buffer	1.5 μl	1.5 μl
PCR fragment DNA*	5.0 μl	–
Plasmid DNA*	5.0 μl	5.0 μl
T4 DNA ligase	1.0 μl	1.0 μl
Total	15 μl	15 μl

* You may wish to verify the efficiency of your gel extraction procedure by quantitating the vector and PCR-amplified fragments prior to use in ligation. Traditionally, equimolar ratios of vector: insert DNA are used to perform a ligation. We have found, however, that intangible variables (i.e. enzymes used, fragment sizes and concentrations, sample purity, etc.) all contribute to the success of a given ligation. Thus precise quantification of the DNA reactants does not necessarily assure success. The volumes suggested for use in this reaction have been determined empirically to be optimal for a wide range of inserts and plasmids in our hands.

(2) Mix the components well. *Do not vortex.* Try not to form bubbles. If bubbles form, quick spin the tube in a microcentrifuge and then put back on ice. Store remaining vector and PCR fragment DNA at −20 °C.

(3) Incubate the reaction overnight at 16 °C. While this temperature is optimal, ligations may also be performed at 4 °C. The reaction is then ready for bacterial transformation. The ligation reaction can be allowed to incubate for up to 2 days, if necessary, then stored at 9 °C.

7.5 Transformation of ligation reactions

The tube containing your final ligation reaction will consist of several different molecules, including a mixture of unligated vector and insert, vector that was ligated without an insert, and a vector ligated with insert DNA (your desired result). The best way to separate and identify these different possibilities is via bacterial transformation. The L4440 plasmid contains the Amp^R gene, which produces the enzyme β-lactamase that breaks down the antibiotic ampicillin. It is this characteristic that will allow you to identify cells that have successfully taken up the circularized plasmid DNA. After a successful transformation, individual bacterial colonies will represent transformants containing circularized vector with or without insert DNA.

You have previously performed a transformation using electrocompetent cells (exercise in Section 5.2). In this lab, you will learn to use chemically competent cells. The procedure below outlines a transformation using the ultracompetent XL10-Gold cells obtained from Stratagene. If you use a different source of cells, either commercially purchased or prepared 'in house', be sure to follow the appropriate transformation protocol provided by the manufacturer. Whatever your source of competent cells, we recommend that they be tested previously for transformation efficiencies above $1 \times 10^7/\mu g$ for optimal ligation success.

Procedural overview

As in the electroporation protocol, it is important to keep the cells on ice unless expressly directed otherwise. Chemically competent cells have been carefully pretreated to ensure high transformation efficiency and therefore should be handled delicately. After the cells have thawed on ice, they are aliquoted into individual pre-chilled snap-cap tubes for each transformation. B-Mercaptoethanol (β-ME) is then added, which functions to increase the transformation efficiency. High transformation efficiency is very important in order to maximize the likelihood of identifying successful ligation products.

After treatment with β-ME, samples from each ligation reaction are added to individual tubes of cells and allowed to incubate on ice for 30 min. This enables the DNA and cells to come into physical proximity with each other. Most chemically competent cells are pretreated with $CaCl_2$, which coats the cells and increases the overall positive charge. This functions in a similar capacity to LiAc used in the yeast transformations of Experiment 3. Following this incubation period, the mixture is heat-shocked for exactly 30 s. This step is very important because it serves to expand cellular boundaries and allows physically adjacent DNA to rush into the cells. Placing the cells instantly on ice after the heat-shock snaps the cells closed, trapping the DNA within the bacteria.

The cells are then allowed to recover for 1 h in rich growth media (SOC). This provides time for the cells to repair any cellular damage from the transformation while beginning to replicate the newly introduced DNA and producing β-lactamase. You will spread the cells onto LB agar plates containing ampicillin to select for transformants. The bacterial plates are incubated overnight to allow transformed cells to grow into colonies. Visible colonies are composed of cells that contain the vector L4440. In the next laboratory exercise, you will use PCR to determine which of those bacteria carry vectors that also contain your inserted PCR fragment DNA.

Reagents

Ligation reactions from previous lab

Chemically competent *E. coli* cells (Stratagene, XL10-GOLD, catalog no. 200314, or equivalent) β-ME, typically provided with cells

SOC broth (1 ml per transformation), prewarmed to 42 °C

LB agar plates +100 μg/ml ampicillin (two per transformation)

70 percent ethanol (for sterilizing cell spreader)

Equipment

Ice bucket

Snap-cap tubes (such as Falcon 2059), sterile (one per transformation)

42 °C Waterbath

37 °C Incubator, shaking

Bunsen burner or ethanol lamp

Cell spreader

37 °C Incubator, stationary

Note: It is critical to the success of the procedure that cells be kept cold

(1) Remove competent cells from −80 °C freezer and thaw on ice.

(2) Place and label the required number of sterilized 15 ml snap-cap tubes on ice (one tube per transformation) corresponding to each ligation performed in the exercise of Section 7.4.3).

(3) Gently mix the cells by inverting the tube and then aliquot 100 μl of cells into the bottom of each chilled snap-cap tube. Add 4 μl of β-ME to each tube of cells and incubate on ice for 10 min, while briefly swirling every 2 min.

(4) Add 5 μl of DNA from each of your ligations directly to tubes containing 100 μl of competent cells, moving the pipette through the cells while dispensing. Gently tap tubes to mix.

(5) Incubate cells on ice for at least 30–60 min (longer is better).

(6) Heat-shock cells for *exactly* 30 s by placing the tubes in a waterbath set to 42 °C. Do not shake.

(7) Immediately transfer the tubes to ice for 2 min.

(8) Sterilely add 900 μl of prewarmed (42 °C) SOC broth to each tube and shake at 225 rpm for 1 h at 37 °C. During this period, label two LB agar plates with ampicillin for each reaction. Label on the bottom.

(9) Using a sterile technique and a cell spreader, spread 200 μl of cells from each transformation onto the appropriately labeled plates. Perform in duplicate if additional plates are available.

(10) Allow liquid to dry into plates. Incubate the plates, inverted, at 37 °C.

(11) Remove the plates from the incubator after 12–16 h to avoid the formation of 'satellite' colonies. Satellites are small colonies resulting from opportunistic bacteria growing within the zone of degraded ampicillin created by true transformants. Store plates at 4 °C.

(12) Compare control plate growth (vector only) with your ligation reaction plate growth. You should see very few colonies on the control plate when compared to the ligation plate. If there are many colonies on the control plate, this indicates that your vector self-ligated easily. It will therefore be more time consuming to identify positive results from your ligation growth plates.

Instructors' notes

Unsuccessful transformation following ligation

If no transformants appear on your plates, you will need to begin again with the restriction enzyme digestion (exercise in Section 7.3.3). Consider doubling the amount of DNA used for both digests in order to increase the final concentration of DNA used in the ligation reaction. You will need to adjust the entire restriction enzyme mixture if you do increase the DNA portion. The DNA sample should not equal more than half your total reaction volume, due to impurities that may inhibit restriction enzymes. You will also need to increase the amount of enzymes used. For 1 μg of DNA, use at least one unit of enzyme. Do not allow the total enzyme volume to equal more than 5 percent of your total reaction volume because excess glycerol (which the enzymes are stored in) can interfere with the digestion. Also, it is possible that the enzymes did not cut well, especially in the case of your insert. You may consider first cloning your PCR reaction into a TA or Topo vector (Invitrogen) before proceeding with the restriction enzyme digest. These are commercially available vectors that will take up any piece of DNA without the use of restriction enzymes or ligation reactions.

You can also allow the ligation reaction to proceed for 2 days to increase the likelihood of a successful ligation reaction.

7.6 PCR screening of transformation colonies

Provided that growth has appeared on the transformation plates from the previous lab exercise, you are ready to proceed with screening of your transformants. It is necessary to distinguish between colonies that have taken up a recombinant vector from transformants that result from a self-ligated vector without any insert DNA. Recircularized vector can be transformed into bacterial cells as well as vector DNA containing an insert.

If you were able to use two different restriction enzymes in your cloning, you should find a low percentage of transformants containing vectors without inserts. This is because the vector ends will not be compatible with each other but rather only to the comparable ends on the insert DNA. However, if the restriction enzymes were not very efficient, or if the digest was allowed to proceed too long so that the enzymes displayed non-specific activity, you will find a large percentage of insert-minus vectors due to blunt-ended self-ligation. It is common to obtain a background of transformants containing self-ligated vector because intramolecular interactions (self-ligation) are highly favored over intermolecular interactions (ligation of vector plus insert) in a given ligation reaction. Therefore, it is always advisable to screen your transformants for the presence of a cloned insert before proceeding with downstream applications.

This exercise outlines a PCR-based screening method that facilitates rapid identification of transformants containing successful recombinants. You will perform a basic PCR reaction using primers homologous to your insert sequence only. If a bacterial colony contains a vector that also has your insert DNA, the primers will bind and yield an appropriately sized PCR product visible on an agarose gel. However, if the vector does not contain the insert DNA, no PCR product will be produced, resulting in no visible band on a gel stained with EtBr.

Procedural overview

The most important thing to pay attention to in this laboratory exercise is labeling! It is critical that you are able to determine unequivocally which bacteria correspond to positive PCR results in order to grow more of those cells for future experimentation. If your tubes are not all labeled accurately, you may incorrectly grow insert-less bacteria, which you would then propagate without ever realizing your mistake. Work with a partner and carefully check each tube as you proceed.

This PCR differs only in set-up and the initial step from the previous PCR that you have performed (Section 7.2). Rather than starting with isolated DNA as your template, you will use a toothpick to carefully transfer cells directly from a bacterial transformation plate into PCR tubes. You will not be able to see the individual cells (do not worry, they are there in the tube), yet this amount of starting material is sufficient for the sensitivity achieved by the PCR. After you have introduced some cells into the PCR mixture, you will then place the same toothpick (which will still retain a small amount of cells) into a tube of growth media. This maintains the cells until you are able to check your PCR reactions for the presence of a positive product.

The first step of the PCR reaction serves to heat the mixture until the cells rupture, releasing the DNA contained within the bacteria. The PCR then proceeds as usual through a series of denaturation, extension, and elongation cycles (see the exercise in Section 7.2 for a refresher on PCR). A portion of the completed PCR reaction resulting from each colony will then be loaded onto an agarose gel to identify potential reaction products. Cells that yield a band on the gel of the appropriate size for the predicted PCR product (size of your insert) can then be placed in a larger volume of media and grown overnight in preparation for plasmid isolation.

Reagents (see Appendix II for recipes)

Double-distilled water, sterile

10X Taq Buffer A (Fisher Scientific)

Taq DNA polymerase (Fisher Scientific), 1 Un/μl stock concentration

dNTP mixture, 10 mM (contains dATP, dCTP, dGTP, and dTTP)

Primers 1 and 2, 20 pmol/μl (as in the original PCR of Section 7.2)

LB broth containing 100 mg/ml ampicillin

Transformation plates from previous lab exercise

Agarose (powder)

1X TAE buffer

Ethidium bromide, 0.625 mg/ml (CLP dropper bottle 5450) *(Careful: it is a mutagen!)*

DNA molecular weight marker, 1 kb ladder (New England Biolabs) containing 1X final concentration gel loading dye

6X gel loading dye

Equipment

Microcentrifuge tubes, 1.5 ml, sterile

Thin-walled PCR tubes, 0.2 ml, sterile

Centrifuge tubes, 15 ml, sterile

Toothpicks, sterile

Ice bucket

Thermocycler, programmed as detailed below

Glass flask with boiling cap

Microwave

Casting trays or laboratory tape for pouring gels for electrophoresis

Parafilm

Gel comb with appropriate number of wells

Gel trays, 9–10 cm size

Gel electrophoresis apparatus

UV illumination source

Gel documentation system (such as a camera attached to UV illumination)

Snap-cap tubes (such as Falcon 2029), sterile

The PCR program

Step	Temp.	Time	Purpose of each step
1	95 °C	10:00	Lysis of bacteria
2	94 °C	0:30	Denaturation of DNA
3	54 °C	1:00	Annealing of primers
4	72 °C	2:00	Extension by polymerase
	Repeat Steps 2–4, 29×		Cycling
5	72 °C	6:00	Final extension

NOTE: To minimize contamination of the PCRs, wear gloves throughout this procedure.

(1) Label a set of 18 0.2-ml PCR tubes and a set of 18 1.5-ml microcentrifuge tubes, respectively.

(2) Place 300 μl of LB + AMP into all 18 of the labeled 1.5-ml microcentrifuge tubes.

(3) Prepare Master Mix (as shown here, it is sufficient for 18 reactions + extra for pipetting error):

10 × Taq buffer	50 µl
Primer 1	10 µl
Primer 2	10 µl
10 mM dNTP	10 µl
Taq DNA polymerase	7 µl
Double-distilled water	413 µl
Total	500 µl

(4) Place 25 µl of Master Mix into all 18 of the 0.2-ml PCR tubes on ice.

(5) Using a sterile technique and toothpicks, with a partner, pick a single colony from the transformation plate. After lightly touching a colony, the toothpick should be twirled into the bottom of the PCR tube and then placed into the correspondingly labeled microcentrifuge tube. *It is very important that you do not take any of the agar from the plate with the toothpick into the PCR tube!* Impurities in agar can inhibit PCR reactions.

(6) Break the toothpick in half so that the end with the bacteria falls into the tube, and close the microcentrifuge tube. Throw away the half of the toothpick that you have touched.

(7) Repeat 17 times with *different* colonies. It is essential to choose well-isolated colonies to avoid introducing multiple templates into a single PCR tube. Be sure to cap each PCR tube tightly.

(8) Place the reactions in a preheated thermocycler and run the PCR program according to the parameters above.

(9) While the program is running, prepare a 0.8 percent agarose gel by dissolving 0.8 g of agarose into 100 ml of 1X TAE buffer (use the microwave). Add two drops of ethidium bromide *(wear gloves!)*. If a refresher is needed regarding gel electrophoresis, refer to the exercise in Section 7.3.1.

(10) When the PCR program ends (~3 h), add 5 µl of 6X Gel loading dye directly to the PCR tubes.

(11) Place gel in electrophoresis chamber and cover with ~1 cm of 1X TAE buffer. Remember to place the wells adjacent to the negative electrode. Mix the PCR samples well by pipeting up and down, then load 6 µl into each lane of the gel.

Load $0.5\,\mu g$ of 1 kb molecular weight ladder into an empty lane, as directed in the manufacturer's instructions. Run at 80–100 V for 45–60 min.

(12) Place the gel on a UV transilluminator and document with a photograph. Examine the photograph to determine which reactions yielded the desired size band.

(13) For reactions that contain your band of interest, take the 1.5-ml tube of media that contains the corresponding toothpick and add it to another 4 ml of LB + AMP in a snap-cap tube for growth overnight shaking at $37°C$, 225 rpm. Even if you will not be performing plasmid isolation the next day, it is important to transfer your positive colonies to fresh growth media immediately after examining your gel, otherwise the cells may die in the microcentrifuge tube.

(14) After the cells are allowed to grow overnight, proceed with the mini-prep procedure to isolate your DNA of interest (next lab).

Instructors notes

Identifying positive transformants

As an alternative to this procedure, it is possible to randomly choose isolated transformant colonies for plasmid purification using the mini-prep procedure and then verifiy the presence of inserts using restriction enzyme digest. If time is a constraining factor, simply skip this laboratory and proceed immediately to the exercises in Section 7.7 and 7.8. It is recommended that each group of students perform a minimum of 24 mini-preps (maximum capacity of a typical microcentrifuge) to increase the chances of identifying a successful ligation product.

7.7 Small-scale isolation of plasmid DNA from *E. coli*: the mini-prep

After you have successfully identified bacterial colonies containing your target insert you will need to proceed with growing more of those bacteria and then isolating purified plasmid DNA for use in RNAi (Chapter 8). Please review Section 5.3 for a refresher on mini-prep methodology before proceeding.

Reagents (see Appendix II for recipes)

LB broth + 100 μg/ml ampicillin

Qiagen QIAprep spin column kit

Equipment

Snap-cap tubes (such as Falcon 2059), sterile

Microcentrifuge tubes, 1.5 ml, sterile

Microcentrifuge

Vortex

Vacuum aspirator (optional)

Before the start of the lab

If you performed the previous exercise, perform Steps 1 and 2 after analyzing the results of your PCR screening via gel electrophoresis.

Day 1 (late afternoon)

(1) Inoculate 3 ml of LB + 50 μg/μl ampicillin with one colony from each positive bacterial clone into a 15-ml snap-cap tube.

(2) Place in a 37 °C waterbath and shake for ~14–16 h at 225 rpm.

Day of the lab (Day 2)

(3) Label the following set of tubes with the name of each culture from which plasmid DNA will be isolated:

(a) 1.5-ml tube with lid;

(b) 2-ml collection tube – place the QIAprep column from the Qiagen kit in the 2-ml tube;

(c) 1.5-ml tube with lid cut off.

(4) Wearing gloves, lightly vortex the culture and carefully pour ~ 1.5 ml of culture into an appropriately labeled 1.5-ml tube (with lid). At this time, place Buffer P1 (stored at 4 °C) on ice for use below.

(5) Spin down bacteria in a microcentrifuge for 1 min at top speed.

(6) Pour off supernatant or remove with a vacuum aspirator. Be careful not to suck up the bacterial pellet if using an aspirator.

(7) Repeat Steps 4–6. Remove as much supernatant as possible.

(8) Completely resuspend pelleted bacterial cells in 250 µl of Buffer P1 by vortexing.

(9) Add 250 µl of Buffer P2 and gently invert the tube *eight times* to mix. Do not vortex!!! Work quickly because the cells should not be in Buffer P2 for more than 5 min.

(10) Add 350 µl of Buffer N3 and invert the tube immediately, but gently, *10 times*. Do not vortex.

(11) Centrifuge for 10 min at top speed in a microcentrifuge.

(12) Carefully pour the supernatant from Step 11 into the QIAprep column that is resting inside the 2-ml collection tube. Be careful not to pour any of the debris into the column.

(13) Centrifuge for 1 min at top speed in a microcentrifuge. Discard the flow-through, but save the collection tube.

(14) Wash the QIAprep spin column by adding 500 µl of Buffer PB.

(15) Centrifuge for 1 min and then discard the flow-through, but save the collection tube.

(16) Wash the QIAprep spin column by adding 750 µl of Buffer PE.

(17) Centrifuging for 1 min at top speed in a microcentrifuge. Discard the flow-through but save the collection tube.

(18) Centrifuge the column once more for 1–2 min at top speed to remove residual wash buffer.

(19) Place the QIAprep column in the 1.5-ml microcentrifuge tube that does not have a lid (the 2-ml tube can be discarded).

(20) Elute the DNA by adding 50 µl of Buffer EB to the center of the QIAprep column. Watch carefully to be sure that the solution covers the silica and does not cling to the sides of the column. Let it stand for 1 min.

(21) Centrifuge for 1 min at top speed in a microcentrifuge.

(22) Discard the QIAprep column and transfer the DNA to a fresh 1.5-ml microcentrifuge tube with a lid. Label your tube appropriately. Store purified plasmid DNA at $-20\,^\circ$C.

(23) For each positive clone isolated using the mini-prep procedure, you should set up the reaction outlined below in the next exercise.

7.8 Verifying successful ligation by restriction digestion

After you have isolated plasmid DNA in the previous exercise it is recommended that you perform a final verification for the presence of your insert. The simplest way this can be done is using a restriction enzyme method, known as a 'check digest'. Following the check digest, you simply run the reaction on an agarose gel and verify that your predicted bands are present. You have then successfully cloned your DNA target and are ready to proceed with the final goal of this course: functional analysis of a gene by double-stranded RNA-mediated interference or RNAi (Experiment 8).

Procedural overview

The volumes outlined are for an individual digest and are very small, typically beneath the accurate range of a pipettor, therefore, set up a Master Mix using the recipe below in which you mix together enough reagents for the number of digests you need to perform plus one extra reaction to account for pipetting errors. The enzymes and buffers used in the check digest correspond to the same ones you used in the restriction enzyme digest prior to ligation (exercise in Section 7.3.3). The idea behind the check digest is simply to 'pop' the insert back out of the vector, thereby verifying insertion of the target DNA.

Reagents

Restriction enzymes 1 and 2, $10\,\mu\text{ml}^{-1}$ (as used in the cloning reaction of Section 7.3)

100X Bovine serum albumen (BSA)

10X Restriction enzyme buffer, appropriate for use with both of your enzymes (buffers appropriate for use with two enzymes can be found in the New England Biolabs catalog or online under the 'Double Digestion' reference appendix or in the NEBuffer Activity Chart)

Agarose (powder)

1X TAE buffer

Ethidium bromide, 0.625 mg/ml (CLP dropper bottle 5450) *(Careful: it is a mutagen!)*

DNA molecular weight marker, 1 kb ladder (New England Biolabs) containing $1 \times$ final concentration gel loading dye

6X Gel loading dye

Equipment

37 °C Waterbath or heating block (or other temperature, if appropriate for your restriction enzymes)

Glass flask with boiling cap

Microwave

Casting trays or laboratory tape for pouring gels for electrophoresis

Parafilm

Gel combs

Gel trays

Gel electrophoresis apparatus

UV illumination source

Gel documentation system (such as a camera attached to UV illumination)

Microcentrifuge tubes, 0.5 ml, sterile

Day 1

(1) Make a Master Mix using the following base reaction components:

10X buffer	2.0 μl
100X BSA	0.2 μl
Double-distilled water	12.4 μl
Enzyme 1	0.2 μl
Enzyme 2	0.2 μl

Set up enough Master Mix for $n + 1$ reactions, where n is the number of transformants you are verifying.

(2) After adding all the reagents and carefully mixing, aliquot 15 µl into individual 0.5-ml centrifuge tubes (sterile). Use one tube for each check digest. These tubes should be labeled to correspond to the transformant being verified.

(3) After aliquoting the digest mix, carefully add 5 µl of each purified plasmid from the previous exercise to the appropriately labeled digest tube, using a fresh pipet tip for each sample.

(4) Incubate the digest overnight at the appropriate reaction temperature, typically 37 °C.

Day 2

(5) Perform electrophoresis on your samples using an agarose gel as outlined in the exercise of Section 7.3.1. Because you will not need to keep the digest samples, you can simply add 3 µl of 6 × gel loading dye directly to each digest tube, mix well, and then load 6 µl into the gel.

(6) Run gel at 80–100 V for 45–60 min.

(7) Follow instructions for use of the specific photography set-up you are using and print a copy of your gel picture. Dispose of the gel in an appropriate waste container. Record data and interpret the success of your check digest reactions.

(8) Analyze the gel using a UV transilluminator and verify that you have two bands: one should correspond to the size of your vector and the other to the size of your insert.

Instructor's notes

If rushed for time, it may be possible to do this exercise in just one longer class period. In this case, a minimal digestion time for the restriction enzyme reaction would be 1 h, during which time students prepare a gel for loading immediately at completion of that time.

8 RNA-mediated interference by bacterial feeding

One of the best ways to determine what a gene or gene product does within a cell is to disrupt its function and examine any resulting defects. In previous years, such investigations were performed by changing a DNA sequence in a relatively permanent fashion. A gene would be altered physically, reintroduced into an organism, and then the phenotypic effects of a specific genetic change would be studied. Alternatively, chemical mutagenesis could be used to randomly mutate sequences in an attempt to find specific classes of effects. The mutated gene then would be identified painstakingly through genetic mapping and studied in further detail.

In 1998, *C. elegans* researchers Andrew Fire and Craig Mello reported their observations that the introduction of double-stranded RNA (dsRNA) results in a specific and dramatic knockdown of the corresponding RNA sequence within the organism. This phenomenon, known as RNA interference (RNAi), is highly potent, with only a few molecules of dsRNA yielding effective and specific knockdown. In *C. elegans*, RNAi results in multiple levels of defects of variable severity, yielding graduated effects such as those observed in an allelic series. Furthermore, knockdown is not restricted to the site of introduction, but rather spreads throughout the organism and thus into F_1 offspring developing within the gonad.

In the subsequent years of study, scientists have analyzed and begun to understand how RNAi works. It serves as a defense against viral infection but may not have evolved that way. What is known is that the introduced dsRNA is recognized in cells by an enzyme complex termed Dicer–RDE. After recognition, Dicer cleaves the dsRNA into short 21–23 bp fragments of dsRNA known as short interfering RNAs (siRNAs), which serve as multiple triggers and allow the RNAi effect to spread rapidly through the organism. The siRNA–Dicer complex then binds with a complex of proteins called the

Integrated Genomics   Guy A. Caldwell, Shelli N. Williams, Kim A. Caldwell
© 2006 John Wiley & Sons, Ltd

RNA-induced silencing complex (RISC). Here the siRNAs are unwound so that they can base-pair with the complementary mRNA expressed naturally within the cells. This enables the RNAi protein complex to specifically target a given transcript within an organism, resulting in degradation of the target mRNA, thereby blocking production of the corresponding protein and, in some cases, yielding a recognizable phenotype.

Initial uses of RNAi in worms involved *in vitro* synthesis of dsRNA and manual microinjection into the gonad of parental animals. However, further studies revealed an interesting phenomenon. As you know, *C. elegans* eat bacteria. It was soon discovered that bacterial strains could be used to synthesize dsRNA *in vivo* using a DNA template corresponding to a target gene. *In vivo* synthesis requires a specially engineered strain of bacteria expressing the T7 polymerase (which transcribes RNA from DNA) into which the target DNA template has been introduced as part of a plasmid via transformation. The DNA template within the plasmid is flanked by T7 promoters, so that the polymerase synthesizes both strands at once, resulting in dsRNA specific to the target. The bacterial strain used by *C. elegans* researchers is discussed in greater detail in the exercise of Section 8.1. The transformed bacteria are fed to *C. elegans* and when the bacterial cells are digested the dsRNA is released into the body of the worm. As mentioned above, because dsRNA is able to cross cellular boundaries, the RNAi effect quickly moves out of the intestine and throughout the animal, resulting in specific knockdown of the target transcript in the whole organism and the progeny of hermaphrodites.

When performing an RNAi feeding experiment, careful planning is critical. For example, to maximize the efficiency of RNAi, you should always use fresh reagents in each feeding experiment. This includes all plates, bacteria strains, cultures, and phenotypic analysis reagents (such as plates supplemented with chemicals). A basic outline is provided below. We recommend that each step of an RNAi experiment be performed immediately after the preceding, with no intervening wait days. Please note that the days listed below correspond to days in the overall RNAi experiment. Individual exercises may have their own 'Day 1', etc., listed within the experimental procedure.

- Day 1: media preparation for RNAi feeding (exercise in Section 8.2)

- Day 2: transformation of RNAi feeding strain HT115(DE3) (exercise in Section 8.3)

- Day 3: RNA interference by bacterial feeding of *C. elegans* (exercise in Section 8.4)

- Day 4: continue the exercise in Section 8.4

- Day 5: continue the exercise in Section 8.4

- Day 6: continue the exercise in Section 8.4

- Day n: analyzing effects of dsRNAi (exercise in Section 8.5)

Additionally, numerous factors can affect the success of an RNAi analysis, including age and genotype of the worms, amount of inducing chemical (see Section 8.2), length of exposure to dsRNA-producing bacteria, and of course the specific target gene. *Please refer to the 'Instructor's notes' before setting up your experiment. It is important that you examine all the protocols before beginning so that you can make a specific schedule for use before embarking on a round of RNAi-based analysis.*

Alternatively, your instructor may have acquired a specific RNAi bacterial strain from another source. If this is the case, proceed directly to the exercise in Section 8.4. You should follow the instructions received with your RNAi clone for propagating the bacteria.

8.1 Preparation of RNAi-feeding bacteria for transformation

In the previous chapter you cloned your target DNA sequence into plasmid L4440. This target sequence is specific for the gene you wish to knock down using RNAi. In order to study the functional consequence(s) of knocking down this gene in *C. elegans*, you must introduce it into a special strain of bacteria: HT115 (DE3).

The pivotal idea behind RNAi feeding in *C. elegans* is that bacteria can be made to produce the dsRNA needed to trigger targeted knockdown. Worms, in turn, will eat these bacteria, thereby ingesting the dsRNA. However, there is no *E. coli* strain that naturally produces dsRNA, which is recognized as foreign by bacteria and is normally degraded by the enzyme RNase III. Fortunately for *C. elegans* researchers, bacterial strain HT115 was engineered so that it is unable to degrade dsRNA. Scientists used a transposable element (a mobile piece of DNA) to interrupt the RNase III gene in a host strain of *E. coli*. This transposon contains a Tet^R gene (TetB) that conveys tetracycline resistance by transporting this antibiotic out of the cells, serving as a selectable marker just as an antibiotic resistance gene on a plasmid does.

A second key ingredient for a successful RNAi feeding experiment involves the ability of the bacterial strain to produce dsRNA. RNA can be transcribed from plasmid DNA using the T7 polymerase, which is a very active polymerase originally identified in bacteriophage. Strain HT115 was modified further by researchers to produce the T7 polymerase by incorporation of the gene encoding this enzyme directly into the bacterial genome. This was performed using lysogenic viral sequences that naturally incorporate DNA into a host genome. In the ordinary viral lifecycle, the virus would eventually exit the genome, kill and lyse the host cell, and be released into the environment. However, the lysogenic sequences used to incorporate the T7 polymerase into HT115 have mutated and are thus unable to leave the bacterial genome. At this point, the strain is now known as HT115(DE3). The T7 polymerase works by recognizing short DNA sequences that act as promoters of transcription known succinctly as T7 binding sites that flank the MCS of the L4440 vector.

Before beginning your RNAi feeding experiment, you need to prepare a stock of HT115(DE3) cells that are ready (competent) for transformation. You have previously performed transformation by electroporation (Section 5.2) and chemical means (Section 7.5). The HT115(DE3) strain is chemically competent, but you will prepare the cells for transformation rather than purchasing them from a commercial supplier. A stock of HT115(DE3) bacteria can be obtained from the *Caenorhabditis* Genetics Center (CGC).

Procedural overview

Preparing chemically competent cells is relatively easier than preparing electrocompetent cells as you did in the exercise of Section 5.1. The most obvious difference between these procedures is that glassware and plastics used to make chemically competent cells do not have to be acid-washed. Everything must be sterilized but you do not have to worry about removing excess ions from the cellular environment. Please note, however, that cells treated with ionic buffers are fragile and should be kept on ice at all times!

After growing an overnight sample of HT115(DE3) cells to saturation, you will add this starter culture to a larger flask of media. The cells will be allowed to grow to a dense yet still replicating population. This mixture is then centrifuged to collect the cells and resuspended in $CaCl_2$. Incubating the cells on ice with this solution coats the cellular membranes with the polar, positively charged $CaCl_2$ ions. During transformation (Section 8.2), these pretreated cells will attract the negatively charged DNA.

Following incubation with $CaCl_2$, cells are collected by centrifugation and resuspended in a smaller volume to concentrate the cells. Cells are then aliquoted into single-use microcentrifuge tubes and rapidly frozen in a dry ice–ethanol bath, maintaining their health during long-term storage at $-80\,°C$.

Reagents

HT115(DE3) cells, grown on an LB agar plate $+12.5\,\mu g/ml$ tetracycline (Tet)

LB broth $+12.5\,\mu g/ml$ Tet

$50\,mM$ $CaCl_2$, sterile

50 percent glycerol, sterile

Dry ice

95 percent ethanol

Equipment

37 °C Shaking incubator

Centrifuge capable of reaching 1500 g and cooling to 4 °C (e.g. IEC CentraMP4R)

Toothpicks or inoculating loop, sterile

Snap-cap tube, sterile (Falcon 2029)

Centrifuge tubes, 50 ml, sterile (Falcon 2098)

Pyrex baking dish for dry ice – ethanol bath

Microcentrifuge tubes, 0.5 ml, sterile

Ultralow freezer (−80 °C) or liquid nitrogen storage tank

NOTE: Tetracycline is light sensitive. Your tubes should be wrapped in foil unless your incubator is entirely enclosed with no glass viewing windows.

Before the start of the lab (~16–18 h before Step 2)

(1) Inoculate 5 ml of LB broth +12.5 μg/ml Tet with a single colony of *E. coli* strain HT115(DE3). Grow overnight at 37 °C in a shaking incubator at 225 rpm.

Day of the lab

(2) Inoculate 25 ml of LB broth +12.5 μg/ml Tet in a 50-ml sterile centrifuge tube with 250 μl of overnight of culture from Step 1. Grow to an OD_{600} of 0.4–0.5 (~2–3 h).

(3) Centrifuge cells in the same centrifuge tube used for growth at 1500 g for 10 min at 4 °C.

(4) Discard supernatant and place cells immediately on ice.

(5) Gently (no vortexing) resuspend cells in 12.5 ml of cold 50 mM $CaCl_2$; use a sterile glass pipet for resuspending.

(6) Incubate cells on ice for 30 min.

(7) Centrifuge cells as in Step 3.

(8) Discard supernatant and place cells on ice.

(9) Resuspend cells in 2.5 ml of cold 50 mM $CaCl_2$. Keep cells on ice.

(10) Wearing gloves, line up 0.5-ml sterile microcentrifuge tubes in an ice bucket (40–50 tubes) with caps open. Prepare a dry ice – ethanol bath by filling a Pyrex glass dish about one-third full with 95 percent ethanol and mixing in dry ice until the solution is viscous. Place a microcentrifuge tube floater in the glass dish. Nearby, prepare a microcentrifuge tube storage box by placing it in an ice bucket on some dry ice to ensure that it remains cold until final storage at $-80\,°C$.

(11) Working with a partner, dispense 50 μl of cells into a microcentrifuge tube, immediately cap the tube, shake cells to bottom of tube with a rapid downward flick-of-the wrist, and then quickly place the tube in the float in the dry ice – ethanol bath. The cells will freeze in a few seconds. Remove tubes from bath, wipe off excess ethanol with tissue (use a fresh every five tubes), and place into the storage container on dry ice. Immediately repeat until all the cells have been aliquoted and frozen.

(12) When all tubes have been frozen, transfer container to an ultralow freezer ($-80\,°C$) or liquid nitrogen tank.

Bioinformatics exercise

When performing an experiment, many scientists rely on protocols published previously in scientific research papers. Often, small variables can contribute to a significant difference in experimental results, and RNAi feeding in *C. elegans* is no different.

(1) There are references for several different large-scale RNAi screens in *C. elegans* in the Section 8.5. Obtain copies of three of these papers or find three of your own by searching Pubmed for '*C. elegans* RNAi screen'.

(2) Carefully review the Materials and Methods Section of each paper for the procedural outline detailing how dsRNA was delivered to the animals.

(3) Outline major differences and similarities among the procedures.

(4) Keep these papers for use in the Bioinformatics exercises throughout this chapter.

Instructor's notes

It is recommended that you inoculate your overnight culture in Step 1 from a freshly grown culture streaked onto an LB agar plate supplemented with tetracycline. To make an LB + Tet plate, simply follow the recipe in Appendix II; make only a small amount of media (100 ml) so as not to waste many reagents. Make this plate at least one but no more than three days before you plan to streak your HT115(DE3) stock cells. You should streak from your stock 1–3 days before Step 1 of this exercise. After streaking your cells, allow them to grow for 15–18 h at 37 °C. If you streak several days in advance, store your plate at 4 °C wrapped in foil after this overnight growth.

   If you do not have access to a CentraMP4R centrifuge or equivalent model, you can use a Sorvall floor centrifuge equipped with an SS-34 rotor or equivalent model. You will, however, need to sterilize conical-bottom Oak Ridge tubes (50 ml) before the laboratory exercise and transfer the cells from Step 2 into these tubes before proceeding. All other steps will proceed as outlined.

8.2 Media preparation for RNAi feeding

There is another feature of the HT115(DE3) cells used in RNAi feeding that should be discussed. The gene for the bacteriophage T7 DNA polymerase was engineered into this strain at a site adjacent to the *lac* operon. You have previously worked with the *lac* operon, when you performed β-galactosidase staining of worms (exercise in Section 2.1) and yeast (exercise in Section 3.3). As you know, the *lacZ* gene encodes the enzyme β-galactosidase, which normally cleaves lactose into glucose and galactose. This enzyme is also capable of cleaving synthetic analogs such as X-gal. However, it is not this capability that is employed in an RNAi experiment. Instead, here the *lac* operon is exploited to provide inducible control of the T7 polymerase, which will drive dsRNA production. The presence of isopropyl-β-D-thiogalactoside (IPTG) in bacterial growth media is enough to trigger activation of the *lac* operon; this occurs as IPTG binds to a *lacI* repressor protein that normally regulates *lacZ* gene expression, thus the addition of IPTG serves to block repression of genes transcribed by this operon. Because the T7 polymerase is engineered adjacent to the *lac* operon, the addition of increasing levels of IPTG leads to a concomitant increase in T7 polymerase transcription. In all, this system provides for facile regulation of RNAi and is advantageous for controlled dsRNA production (see Figure 8.1).

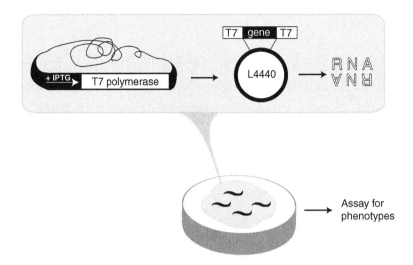

Figure 8.1   Induction of dsRNA production within bacterial cells HT115(DE3)

Procedural overview

The media used in an RNAi feeding experiment are slightly different from those normally used in worm maintenance. Although the basic ingredients utilized are unchanged, the bacteria used in an RNAi feeding experiment carry the RNAi target on a plasmid encoding for ampicillin resistance, therefore instead of supplementing plates with streptomycin, as is typical for worm media, you will instead add ampicillin to your media mixture. The other change required for RNAi plates is the addition of IPTG. The level of IPTG added to the plate corresponds to the level of dsRNA produced by the bacteria. This, in turn, corresponds to the strength of the RNAi knockdown effect. In general, more IPTG results in higher production of dsRNA, resulting in increased transcript knockdown and more severe phenotypic effects. Of course, the IPTG effect can become saturated, wherein increasing the concentration of this inducer has no added effect on dsRNA production. The recipe below calls for the preparation of two different sets of RNAi plates: one set contains a relatively high level of IPTG; the other contains a lower level and should result in less severe phenotypic effects. In this regard, be very careful to label your plates properly.

Reagents (see Appendix II for recipes)

NaCl

Peptone

Agar

5 mg/ml Cholesterol, filter-sterilized

1 M CaCl$_2$, sterile

1 M MgCl$_2$, sterile

1 M Potassium phosphate, sterile

100 mg/ml Ampicillin, filter-sterilized

1 M IPTG, filter-sterilized

Equipment

150-ml Glass beaker

Stirbars (two), 2–2.5 cm

Magnetic stir plate capable of heating

125-ml Pyrex flasks (two)

Aluminum foil

Autoclave

60 °C Waterbath

60 mm × 15 mm Plastic dishes, sterile

Sterile hood (if available, a tissue culture hood)

(1) Add 97 ml of double-distilled water to a 150-ml beaker.

(2) Add the following components to the water:

NaCl	0.3 g
Peptone	0.25 g
Agar	1.7 g

(3) Mix well using a magnetic stirbar and plate. Heat the solution to melt the agar completely. Do *not* allow the solution to boil.

(4) When the components are all in solution, divide the mixture in half. Place each half into a 125-ml flask.

(5) Cap each flask with aluminum foil and autoclave for 20 min on liquid cycle.

(6) When the autoclave is finished, remove the flasks and place one on a stir plate and the second in a waterbath. Stir the first flask slowly. Do *not* allow bubbles to form in the mixture. You should not heat the flask at this stage.

(7) After 15 min, use a pipettor to add *aseptically* the following components in the order listed from the stock solutions:

Cholesterol (5 mg/ml)	50 µl
1 M $CaCl_2$	50 µl
1 M $MgSO_4$	50 µl
1 M Potassium phosphate	1.25 ml
Ampicillin (100 mg/ml)	50 µl
1 M IPTG	50 µl

(8) Allow the media to stir for 2 min. During this time, label six 60 mm × 15 mm sterile plastic dishes (size used for standard worm maintenance) '1 mM'.

(9) *Using a sterile technique*, carefully pour the media into the pre-labeled plates. Use a sterile UV-irradiated or tissue culture hood if available.

(10) Repeat Steps 7–9 with the second flask but with the following change: *add only 25 µl of 1 M IPTG to the media*. Label these six plates '0.5 mM' while the media mixture is stirring.

(11) Allow the plates to dry in the sterile hood for 24–48 h. You must use the plates *within 3 days* of preparing them or the IPTG will begin to lose potency.

Bioinformatics exercise

One way in which scientists confirm or interpret their data is by examining the work of others who study the same gene or genetic pathway. Large-scale RNAi screens are becoming an increasingly popular means for obtaining cursory information on many genes in a single organism or biochemical pathway. However, few screens utilize the exact same screening criteria and it is important to determine differences when using others' data to interpret your own.

Using the three papers you examined in the last Bioinformatics exercise, answer the following:

(1) Determine the level at which animals were examined throughout the screen (i.e. adults, embryos, antibody staining, changes in GFP expression, or DIC microscopy). There will likely be several levels of examination.

(2) Find three similar categories of defects for each paper. For example, embryonic lethality, uncoordination, and sterility are commonly identified phenotypes in large RNAi screens. Determine the experimental criteria used to place results in these different categories. How are they similar or different?

(3) Who analyzed animals? In other words, was a computer used to gather information that was then examined by one or more researchers, or was each animal initially examined by a person?

(4) How many replications were used in these papers?

Instructor's notes

If your students are analyzing multiple gene targets by RNAi feeding at one time, you will need to increase the amount of media appropriately. The recipe above is sufficient for a single feeding experiment targeting one gene sequence.

It is also possible to make a stock of plates to which you will later add IPTG immediately before an RNAi feeding experiment. If time constraints prevent students from making plates during a laboratory meeting, see Appendix II for making plates and supplementing with chemicals. Note, *however, that you should still substitute ampicillin for streptomycin in plates that will be used for RNAi feeding.*

8.3 Transformation of RNAi-feeding strain HT115(DE3)

Prior to bacterial RNAi feeding of *C. elegans*, you must first transform the bacterial-feeding strain HT115(DE3) with the plasmid you have generated (Chapter 7) that will produce dsRNA specific to your target gene. In order to produce dsRNA from your plasmid, a target DNA sequence simply needs to be transcribed on both strands simultaneously. As you know from Chapter 7 (see Figure 7.1), the plasmid L4440 contains two T7 promoter sites, one on either side of the multiple cloning site (MCS) you used to insert your target sequence. Proceed with the transformation protocol below if you have cloned your own RNAi-feeding vector. You will use the cells you prepared in the exercise of Section 8.1 for this transformation.

Procedural overview

You are now familiar with both electroporation and chemical transformation of bacterial cells. This protocol is a very simple chemical transformation. As you remember from the exercise of Section 8.1, these competent cells have been coated with $CaCl_2$, increasing the overall positive charge of the cell membranes. This positive charge serves to attract the overall negatively charged plasmid DNA. Cells and DNA are allowed to incubate on ice so that they may come into physical proximity. Following this incubation period, the mixture is heat shocked for exactly 1 min. This step is very important because it serves to expand cellular boundaries and allows physically adjacent DNA to rush into the cells. Placing the cells instantly on ice after the heat-shock snaps the cells closed, trapping the DNA within the bacteria.

The cells are then allowed to recover for 1 h in rich-growth media (SOC). This provides time for the cells to repair any cellular damage from the transformation. The incubation also provides time for the newly introduced DNA to replicate and to produce β-lactamase, which will provide resistance to ampicillin. You will spread the cells onto LB plates containing ampicillin and tetracycline to select for transformants. Tetracycline is necessary to maintain selection of the transposon that interrupts the RNase III gene (see the exercise in Section 8.1). The bacterial plates are incubated overnight to allow transformed cells to grow into colonies. Visible colonies are composed of HT115(DE3) cells that now carry both the transposon and the plasmid L4440 containing your RNAi gene target. These cells will be used to perform an analysis of gene function by subsequent RNAi feeding of *C. elegans*.

Please note that it is better to use fresh plates (no more than 2–3 days old) that are supplemented with tetracycline. Students can make a small batch of plates when they are preparing the worm media for RNAi use or the instructor can prepare plates prior to the lab. See Appendix II for preparing LB plates supplemented with tetracycline and ampicillin.

Reagents (see Appendix II for recipes)

RNAi target Gene Y cloned into vector L4440

Chemically competent *E. coli* cells of strain HT115(DE3) (from the exercise in Section 8.1)

SOC broth, sterile (1 ml per transformation)

70 percent ethanol for sterilizing cell spreader

LB agar plates $+12.5\,\mu\text{g/ml}$ Tet $+100\,\mu\text{g/ml}$ Amp

Equipment

Snap-cap tubes (such as Falcon 2059), sterile (one per transformation)

Ice bucket

37 °C Waterbath

Bunsen burner or ethanol lamp

37 °C Shaking incubator

Cell spreader

37 °C Stationary incubator

NOTE: Before beginning this lab, note that it is CRITICAL to the success of the procedure that the cells are kept cold. Tetracycline is light sensitive so keep your plates covered when not in use.

(1) Remove competent cells from −80 °C; thaw by laying tubes on top of ice (not down in ice).

(2) Thaw RNAi target Gene Y::L4440 vector on the bench.

(3) Label the required number of sterilized snap-cap tubes and then place on ice (one tube per transformation).

(4) Once everything is thawed, pipet 50 μl of cells into a pre-chilled snap-cap tube for each transformation that will be performed (whole aliquot).

(5) Add 1.5 μl of vector (50–200 ng) into the appropriately labeled tube and twirl the tube gently in your fingers. Do *not* shake the cells harshly or mix by pipetting.

(6) Incubate on ice for 30 min.

(7) Heat-shock the cells for *exactly* 1 min by placing the tubes in a waterbath set to 37 °C. Do not shake.

(8) Immediately transfer the tubes to the ice for 2 min.

(9) *Aseptically* add 1 ml of room-temperature SOC to the tube.

(10) Incubate the cells for 1 h, shaking at 225 rpm in the 37 °C incubator. During this step, label the bottom of your LB + Tet + Amp plates appropriately with the gene target name.

(11) Using a sterile technique, spread 100 μl onto a pre-labeled LB agar plate containing tetracycline and ampicillin. Cover plates immediately when you are finished.

(12) Allow liquid to dry on plates. Incubate the plates, inverted, overnight in the dark at 37 °C.

(13) Remove the plates from the incubator after 12–16 h to avoid the formation of 'satellite' colonies.

Bioinformatics exercise

Now you have experience examining a paper for procedural details. The next important skill for a scientist is the ability to critically examine the data presented within a paper.

(1) Using your three RNAi screen papers, find one gene that falls into the same basic category in all the papers and one that falls into a different category in at least one of the papers.

(2) Are these results truly the same or truly different, based on what you have determined previously about the experimental criteria used in each paper?

(3) Can these apparent differences be reconciled? In other words, spend a little time doing background research on the gene that gave different results in the different screens. Do the data presented in each correlate with data found by more detailed examinations of the gene or gene product?

8.4 RNA interference by bacterial feeding of C. elegans

You are now ready to begin the culminating experiment for this course. As you have progressed through this manual, you have sought to investigate a specific gene product (Gene X) by searching for its interacting partners. You have isolated and identified interactors (Experiments 4–6) and have prepared to examine one or more in greater detail (Experiment 7). The time has come to determine the phenotypic consequences of knocking down transcript levels of those interactors.

This lab details the steps for exposing worms to dsRNA-producing bacteria. You will use the worm plates that you made in the exercise of Section 8.2 and the bacterial strain you transformed in the previous exercise. Each step below is carefully outlined; please follow the timeline as closely as possible. However, be very sure to record precisely in your notebook all details associated with your RNAi experiment. This is a multiple step exercise and will require careful planning and diligent attention on your part. Please review the 'Instructor's notes' for changes that you can choose to make in your RNAi experiment to vary conditions of specific transcript knockdown for different gene targets.

Procedural overview

You begin this laboratory by stripping the worms to which you will feed your dsRNA-producing bacteria. This step is important because you wish to initiate the RNAi effect in healthy animals that will soon begin producing the next generation. Remember, you will actually be examining the offspring of these animals, or the F_1 generation. The animals that you feed are the parental (P_0) generation. Generally, with RNAi feeding it is more difficult to observe phenotypes in the P_0 generation. Because dsRNA can spread easily from the intestine to the developing embryos within the gonad, exposing L4 animals is a good stage to start with for an initial RNAi assay.

The next step in an RNAi experiment is growing the dsRNA-producing bacteria. Remember, you either transformed HT115(DE3) cells in the previous laboratory with the plasmid L4440 containing your target gene sequence or your instructor obtained the strain from an outside source. Regardless, you should inoculate your overnight cultures using bacterial colonies that were grown on LB agar plates supplemented with ampicillin and tetracycline. This ensures that the starting cultures are healthy. You may notice that the overnight culture media is not supplemented with tetracycline. The transposon that conveys tetracycline resistance to HT115(DE3) is very stable; in the length of time you will be growing these cells there is little chance they will spontaneously lose this transposon, therefore tetracycline is not added to this culture; ampicillin is necessary to maintain the non-integrated, extrachromosomal plasmid L4440.

You begin the third day (on the schedule below; may differ according to when you stripped your worms) by plating bacteria from the overnight cultures onto the worm plates you made in Section 8.2. If you followed the directions in this exercise you should have two sets of approximately six plates each, with one set containing a higher concentration of IPTG. You will divide these sets into three plates of each concentration labeled 'A' and three plates of each concentration labeled 'B'.

The reason for this is simple. When you perform an RNAi feeding experiment, it takes some time for the dsRNA to move into the gonad from the intestine. If you left the P_0 animals on a single plate the entire time, the resulting F_1 generation would be a mixed population, some of which had little or no exposure to dsRNA and others that had received a large dose. This protocol calls for you to transfer the P_0 worms from one plate to another after an initial exposure to the RNAi bacteria. Thus you will be able to separate the early F_1 animals, which will likely show little RNAi effect, from the later F_1 animals, which will show stronger phenotypic effects of target transcript knockdown.

Two days after placing your P_0 animals on the 'A' plates you will transfer them to the 'B' plates to provide extended exposure to dsRNA. The worms continue to lay eggs, which will develop into the F_1 animals that you will analyze in the exercise of Section 8.5. It is recommended that you keep your 'A' plates. Remember, you want all reagents to be as fresh as possible during an RNAi experiment. You will allow your second set or 'B' plates to grow for longer (do not let them overgrow). The F_1 animals that develop on the 'A' plates may show less severe RNAi phenotypes associated with knockdown of your target. At 25 °C, it will take several days for your F_1 animals to develop. The assays recommended in the exercise of Section 8.5 are performed on adult animals, therefore carefully monitor your worms so that you can analyze them at class time on the appropriate day for RNAi effects. You, or whoever is monitoring your animals (such as the instructor), may notice that they are growing too fast in relation to your scheduled class time. Should this be the case, the plates can be shifted to a lower temperature to slow worm growth. If this occurs, simply record the time and change in temperature in your notebook.

An important point to note before beginning your RNAi experiment concerns minimizing contamination. As the entire idea behind RNAi feeding is to make worms eat bacteria that are producing dsRNA, you want the worms to eat *only* those bacteria. Contamination is of the utmost concern during every aspect of an RNAi experiment. Wear gloves during all steps. When you are plating cultures, we recommend wiping down your pipettor with ethanol and using freshly autoclaved pipet tips that have never been opened. We also recommend plating overnight cultures in a sterile hood, if available. Wipe down all surfaces, including microscope stages, that will come in contact with RNAi plates with 70 percent ethanol immediately before each step. Any sign of contamination on your plates means that you cannot use those plates for an RNAi-based analysis because there is no way to be sure that your worms ate the correct bacteria.

Reagents (see Appendix II for recipes)

HT115(DE3) cells transformed with L4440 plus your target DNA sequence (exercise in Section 8.3)

LB broth $+100\,\mu g/ml$ ampicillin

NGM worm plates for RNAi supplemented with IPTG (prepared in exercise of Section 8.2)

NGM worm plates seeded with OP50 (control)

70 percent ethanol

Equipment

Snap-cap tubes (such as Falcon 2059), sterile

Sterile toothpicks or inoculating loops

37 °C Shaking incubator

Sterile hood (if available, a tissue culture hood)

Dissecting microscope

Worm pick

Bunsen burner or ethanol lamp

25 °C Incubator

Optional: 20 °C or 15 °C incubator for shifting worm growth temperature

Before the start of the lab

Day 1 (time determined using the lifecycle charts from Experiment 1, Figure 1.1)

(1) Using a bacterial loop, step 3–5 plates of dauer N2 worms onto NGM worm plates seeded with OP50. Place at the appropriate temperature so that you will have L4 animals available on 'Day 3' as listed below. See lifecycle charts in Appendix III, as well.

Day 2 (20–24 h prior to class)

(2) Using a sterile toothpick or inoculating loop, inoculate two snap-cap tubes containing 3 ml of LB broth $+100\,\mu$g/ml ampicillin with one colony of transformed HT115(DE3) containing Gene Y L4440 from a successful RNAi transformation plate (exercise in Section 8.3).

(3) Place in 37 °C waterbath and shake for 15–18 h at 225 rpm.

(4) If you have stored your RNAi plates from the exercise in Section 8.2 at 4 °C, place them on the benchtop oriented with lid up overnight to allow them to dry. This will facilitate absorption of bacterial cultures during the seeding process in Step 7.

Day of the lab: Day 3 (at least 4 h before class begins)

(5) Retrieve the plates made in the exercise of Section 8.2. There should be two sets corresponding to different IPTG concentrations, one labeled '1 mм' and the other labeled '0.5 mм'. Divide each set in half and label three of each set 'A' and the other three of each set 'B'.

(6) Also label three control NGM plates seeded with OP50 'C-A'. These are the control plates.

(7) Carefully pipette 100 μl of bacteria from your overnight cultures onto the plates labeled 'A' for a given IPTG concentration and 175 μl onto the plates labeled 'B'. After the bacteria have been added, carefully swirl the plate by hand to spread the culture into a thinner layer. However, be sure that the bacteria do not touch the side of the plate because worms then may crawl out and dessicate.

(8) Repeat Step 7 for the other set of IPTG-containing plates, pre-labeled 'A' and 'B'.

(9) Allow the plates to dry in a sterile hood or on the benchtop.

At the start of class (~4 h later)

(10) Retrieve the plates that were seeded with 100 μl of bacteria (labeled 'A').

(11) *Using a sterile technique*, carefully add five L4 stage worms onto each 'A' plate. When transferring, it is *very important* that you do *not* carry over excess bacteria from the plate that the worms were growing on to the new RNAi plate. Add another five L4 animals to the control plates that you labeled 'C-A' in Step 6.

(12) Place the plates with worms into a 25 °C incubator.

(13) Allow the plates with 175 μl of bacteria (labeled 'B') to grow overnight at room temperature (for a total 24 h of growth).

Day 4

(14) Transfer the 'B' RNAi plates to 4 °C so that the bacteria do not overgrow.

Day 5

(15) Remove the 'B' RNAi plates from 4 °C and warm then to room temperature prior to the start of the lab.

(16) Transfer all worms from the original 'A' RNAi plates to new plates of the same IPTG concentration labeled 'B'. You should also transfer worms from the control plates to fresh control plates that you have labeled 'C-B'. In both cases, transfer only the P_0 worms; leave all progeny and eggs that may be present on the 'A' plates.

(17) Place all plates at 25 °C (this includes returning the 'A' plates).

(18) Monitor the age of the F_1 offspring carefully on both the control and the RNAi plates. Proceed with the exercise in Section 8.5 when animals are young adults. Refer to the lifecycle charts (Figure 1.1 and Appendix III) for timing developmental progression. If necessary, shift the growth temperature to slow development of the F_1 animals.

Bioinformatics exercises

You now have the requisite skills to critically evaluate the published data and procedures. Use these skills to find available RNAi information on your RNAi target. As you examine papers and data sets, carefully record the following pieces of information:

(1) Basic experimental outline; in particular, take note of how previous methods may differ from the set-up used in this text.

(2) Any available RNAi data for your gene; record specifically what the data category may represent.

(3) Postulate how your experimental set-up may affect the data you will obtain in the next exercise.

(4) Use this information and your instructor's knowledge to determine which of the assays in the next exercise you will perform. If possible, find other assays through literature searches that you may wish to perform or that may be more informative given your previous knowledge of your target Gene X and its interacting gene product Gene Y (including phenotypes associated with either).

Instructor's notes

To properly carry out this exercise, it may be necessary for either the instructor to perform some steps or for students to briefly visit the lab on days when they would not ordinarily meet for class. Few steps in setting up an RNAi experiment take a large amount of time. Using the guidelines we have provided, plan for more labor-intensive days to fall during normal class periods.

We highly recommend that you use the protocol in Appendix III to decontaminate your worm stocks prior to using them in an RNAi feeding experiment. As noted in the experimental outline, contamination is absolutely unacceptable in a bacterial feeding experiment. Using freshly decontaminated worm stocks can prevent the transfer of extraneous bacteria or fungi from the original plate used for worm propagation.

When performing an RNAi experiment, there are numerous factors that you need to consider. For example, an RNAi effect can be titrated to study both embryonic and adult phenotypes. Even though a gene may be essential for early embryonic development, you can decrease the level of IPTG to reduce the RNAi effect, thereby increasing the number of survivors that bypass embryonic lethality (escapers). This enables you to study adult phenotypes caused by reduction in the target transcript. The highest recommended IPTG concentration is 1 mM; you may decrease this to as low as 0.01 mM. Note that while *E-coli* OP5O is included as a negative control, individual HT115(DE3), without IPTG in the media, can also be used.

Another factor to consider is the age of the P_0 generation when you first expose them to the bacteria producing dsRNA. Although L4 animals are typically used as a starting stage for RNAi knockdown, you can choose to expose dauer animals for longer lived transcripts or young adult animals (gonad already fully formed) for targeting of the transcripts required very early in embryonic development.

However, RNAi is not equally efficient in all cell types; in particular, *C. elegans* neurons have been found to be generally resistant to an RNAi effect. Several strains have been isolated that may overcome this problem. The strain *rrf-3* is generally found to be more sensitive to RNAi in all cell types; likewise, additional strains available from the CGC exhibit differential sensitivity or resistance to RNAi. For example, *sid-1* is resistant to systemic RNAi but sensitive to autosomal RNAi (RNAi effect will not spread and thus cannot be used for feeding). Strain *rrf-1* is resistant to RNAi in somatic cells. Furthermore, more sophisticated RNAi experiments can be performed using mutant strains. Such an experiment would be similar to creating a double mutant strain through genetic methods. This would allow you to assess the possible associations between gene products. You may wish to have students set up one set of RNAi experiments using wildtype worms and another set using a mutant or RNAi-sensitive strain. If a strain is available with a mutation in Gene X, studying the effects of RNAi knockdown of Gene Y in this strain would be ideal. You could also choose to have students study the effects of temperature on the RNAi phenotype. Rather than using feeding plates with different concentrations of IPTG, instruct students to make a single concentration but grow the worms being fed dsRNA at two different temperatures (i.e. 20 °C vs. 25 °C). Be sure that your schedule will allow for two different days of analysis if you choose this procedural deviation.

8.5 Analyzing effects of dsRNAi

In this exercise you will analyze the functional consequences of knocking down your target transcript (from previous exercise). Because each target can give a range of phenotypes, we have outlined a variety of simple assays that you can perform on the F_1 offspring of the animals that have been fed dsRNA. Other classes of phenotypes to look for can be found by reading publications describing the large RNAi screens that have been performed in *C. elegans* (Piano *et al.*, 2001; Kamath *et al.*, 2003; Simmer *et al.*, 2003; Sonnichsen *et al.*, 2005). The existing reported data for given genes can also be found in Wormbase, as can all genes yielding a given phenotype by RNAi. Be sure before proceeding with these assays that you have performed the bioinformatics exercise from Section 8.4 so you will have an idea of the type of phenotypes you can expect to see in your RNAi animals. However, because these were large-scale screens, you may observe more subtle phenotypes in your assays. Additionally, the feeding methodology employed in this text differs from the exact procedures used in the large-scale RNAi screens. Small changes in procedural set-up can also reveal additional phenotypes.

Based on what you know about Gene X and the putative interacting gene product, Gene Y, develop a hypothesis with respect to what cellular mechanism you anticipate might be perturbed by RNAi. Use this hypothesis to direct which of the assays listed below might uncover potential phenotypic effects related to the knockdown of your gene. For example, if your original Gene X was known to be expressed in the vulval muscles and/or mutations in this gene yield egg-laying (Egl) phenotypes, then you might anticipate an interacting partner to exhibit this same defect upon RNAi treatment. Thus you would perform a simple gross observation for accumulation of eggs within the body (Egl) as noted in Section 8.5.3. Likewise, if your original or interacting gene product is predicted to function as a component of the cytoskeleton, RNAi might be expected to yield significant embryonic lethality, visible as an overall reduction in the number of offspring on your RNAi plates as outlined in Section 8.5.1.

Importantly, due to the inherent variability in RNAi feeding, even inefficient knockdown of an essential transcript can yield post-embryonic defects, i.e. neuromuscular uncoordination (Unc or thrashing defective worms). Of course, a gene of completely unknown function may yield any number of phenotypes. In this instance, it will be important to prioritize which assays to perform based on the time available. It is recommended that you assay animals from the 'B' plates initially. If time allows for assaying 'A' animals, as you may be more likely to observe post-embryonic effects in these animals.

Please note that some of the following assays require the worms under study to be exposed to different chemicals for several minutes, some as long as 30–60 min. If you will be performing any of these assays, plan each assay so that you can perform several at one time. In other words, if worms need to be on a chemical plate for 30 min, place these animals on the appropriate plate first. During that time, proceed with shorter assays until the 30 min have passed. Then return to the drug plates and complete that assay. Time management is an important part of success in science.

Reagents (see Appendix II for recipes)

Control plates 'C-A' and 'C-B' from the exercise in Section 8.4; worms should be adults (preferably young adults)

RNAi plates 'A' and 'B' from the exercise in Section 8.4; worms should be adults (preferably young adults)

Plates needed for the appropriate assays: please refer to the 'Media preparation' section in Appendix II for information on preparing worm plates supplemented with chemicals as appropriate for the specific assays below

Equipment

Dissecting microscope, bottom illumination

Worm pick

Bunsen burner or ethanol lamp

Toothpick with eyebrow hair glued to the end (see Section 1.2)

8.5.1 Assaying for sterility (Ste) or embryonic lethality (Emb)

Knockdown of genes that affect development of the embryo will most likely result in a reduction in the total number of offspring from the P_0 animals. This is assayed quite simply by counting the overall number of animals on your RNAi plates versus control plates.

(1) Carefully count and record the number of F_1 ('A' and 'B' plates, adults) or F_2 ('A' and 'B' plates, younger larvae) offspring on each of your control plates.

(2) Repeat Step 1 for the animals on your RNAi plates. Do not forget – you have two different concentrations of IPTG that should be analyzed separately.

(3) Examine the number of worms. Is there a difference? A reduction in overall numbers of hatched larvae on your RNAi plates indicates an embryonic requirement for your

target gene. Unhatched eggs will persist on plates for >24 h after removing the parents. Alternatively, gene knockdown can result in sterility (Ste) wherein eggs are not even laid by hermaphrodites.

(4) Form a hypothesis to explain the possible role that your target gene may play in embryonic or gonadal development.

8.5.2 Assaying for growth effect

Knocking down a gene that is required for development of the animal beyond the embryo can result in a change in the average age of F_1 RNAi animals compared with control animals, therefore, to assay for this phenotypic class, simply examine all offspring visible on the plate. Note the average ages of the animals (use the lifecycle images to remind you of the hallmarks of each stage) on your RNAi plates compared with the controls.

(1) Carefully record the approximate ages of the animals on your control plates; a rough division of animals into one or more age groups is sufficient for an initial analysis. This can be accomplished readily by classifying animals as young larvae (L1/L2), older larvae (L3/L4), or adults.

(2) Repeat for your RNAi plates. Do not forget to analyze each concentration of IPTG separately.

(3) How do the average ages of the animals compare? If the RNAi animals are of younger average age, your target gene may be required for a particular developmental process. Speculate on any possible links between your target gene's function and development of the worm.

8.5.3 Assaying for morphological effects

Many gene products are required for the development of a particular cell or tissue type. Below are a few of the easily observed classes of morphological phenotypes that may result from RNAi. See Appendix IV for a more detailed list of potential phenotypes and a brief description of each classification. See Figure 8.2 for examples of some common phenotypes.

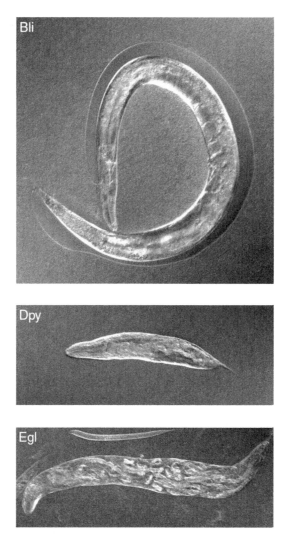

Figure 8.2 Examples of three different types of *Caenorhabditis elegans* mutant phentoypes

(1) Carefully examine your RNAi animals. Are there any obvious physiological defects? Some commonly observed defects are:

    (a) Blistered cuticle (Bli)

    (b) Dumpy (Dpy)

    (c) Abnormal eversion of vulva (Evl)

    (d) Egg-laying defective (Egl)

(2) Based upon any defects observed, hypothesize how your target gene may contribute to the development of this aberrant morphology.

8.5.4 Assaying for general neuromuscular effects

Although it has been generally observed that neurons are more resistant to RNAi than other cell types, this is not always the case and varies with the neuron type and target gene analyzed. Outlined below are assays for examining different classes of neuromuscular effects in *C. elegans*.

Uncoordination

Neuromuscular defects may be the result of affecting a transcript required by either neurons or the muscles they innervate. A general assay for determining if your target may be required in these cell types is to examine the movement patterns of your RNAi animals. You could then perform a more fine-tuned assay to determine which cell type has been affected, as described in later sections.

(1) Carefully place a single control worm onto a freshly seeded plate that has no tracks in the bacterial lawn. Allow the worm to crawl through the plate for 5 min. Remove the worm after the 5 min have passed to prevent excess tracking through the lawn.

(2) Repeat with four more control animals. *Use a fresh plate each time!!*

(3) Repeat Steps 1 and 2 with worms from each of your IPTG-induced sets of RNAi experiments.

(4) Examine the pattern of tracks left by the worms on each plate. Do the RNAi animals leave tracks that look the same as the control animals? Is there a change in the amplitude (height of curve, measured from the center) or wavelength (length from the top of one curve to the top of the next) of each body bend? If so, record how they differ in your notebook.

(5) Based upon what you know about your target Gene Y, can you hypothesize whether you are affecting neuronal or muscular cell types?

Thrashing

Another type of general neuromuscular assay is the thrashing assay. During this assay you are examining a worm's fast response, which differs on the molecular level from the mechanisms responsible for movement on a plate.

(1) Place a single worm from a control plate into a 30-μl drop of M9 buffer on a glass slide. Allow the worm to acclimatize to this new environment for 1 min.

(2) Carefully observe the worm for 2 min. Count the number of head thrashes (number of anterior body bends, encompassing movement from left to right) you observe in that time period. A wildtype worm will thrash an average of 150 times per minute.

(3) Repeat Steps 1 and 2 four more times with control animals. Record the results for each animal and then average the results.

(4) Repeat Steps 1–3 with five RNAi animals. Again, record individual results and also the average number of thrashes.

(5) Again, hypothesize as to whether RNAi knockdown of your gene has affected neuronal or muscular cell types.

Defecation

In *C. elegans*, expulsion of waste occurs through a set pattern of four distinct movements approximately every 42 s. First the posterior end of the animal will contract, followed approximately 1 s later by relaxation. The anterior end contracts next and finally waste is expelled through the anus. The timing of the defecation cycle is invariant, regardless of temperature.

(1) Carefully observe a single control animal under magnification strong enough to see the four defecation movements. Record the time between 3–5 defecation cycles for this animal. Take careful note of the defecation movements to be sure that each is occurring properly. Review the original defecation paper (Thomas, 1990) for further clarification if necessary.

(2) Repeat Step 1 for four more control animals. Record individual times for each animal and calculate the average time between defecation for all five animals.

(3) Repeat Steps 1 and 2 with animals from your RNAi plates. Again, record both individual results and the average length of the defecation cycle.

(4) Form a hypothesis regarding the possible effect of RNAi knockdown of your target upon the defecation cycle and movement pattern.

8.5.5 Assaying for specific neuronal effects

It is also possible to assay for defects in specific classes of neurons within the worm. Again, listed below are only a few possible assays. Search Wormbase or the worm literature for additional neuronal classes or assays that may be more specific to your target gene.

Aldicarb sensitivity (acetylcholine-specific)

Aldicarb is a chemical that inhibits acetylcholinesterase, an enzyme that breaks down acetylcholine in the synaptic cleft following release of this neurotransmitter. In *C. elegans*, aldicarb serves as a simple means to test for defects in exocytosis of this common neurotransmitter. Following aldicarb exposure, wildtype worms experience hypercontraction of the muscles and eventually die. Research has shown that even relatively modest changes to the acetylcholine system can alter aldicarb response. Knocking down a transcript needed for proper function of the acetylcholine system may lead to either hypersensitivity or resistance to aldicarb.

(1) Obtain an NGM worm plate supplemented with 0.5 mM aldicarb from your instructor.

(2) Place 10–15 worms from your control plates onto this plate.

(3) Allow the worms to stay on the plates for 30–60 min.

(4) Examine the plate: count the number of worms that are paralyzed. Paralyzed worms will not respond to prodding with a worm pick or to tapping the plate.

(5) Repeat with worms from each of your RNAi experiments, using a new aldicarb plate each time.

(6) Using your knowledge of Gene Y, hypothesize why you may have observed aldicarb-related defects in the RNAi worms.

Osmosensation assay (head-sensory-neuron-specific)

Worms ordinarily avoid areas of high osmotic strength; when encountering such an area they will quickly back up (several body bends) and then turn to crawl away. This behavior is controlled by several neurons located in the head of the animal. Analyzing for defects associated with the neurons controlling this behavior is relatively simple, because you need only test whether or not the worms avoid a barrier of high osmolarity.

(1) Obtain an NGM plate that is not seeded from your instructor. Also obtain a solution of 4 M NaCl containing the indicator phenol-red (this non-toxic dye allows you to see the solution).

(2) Carefully dip the cap of an extra-fine Sharpie or other marker (radius of 1 cm) into the 4 M NaCl solution with dye. Do not saturate the cap. You just want the bottom to be wet.

(3) Gently place the dampened cap onto the center of an unseeded worm plate and lift without rubbing it around. You should see a ring of liquid on the plate that was left from the cap.

(4) Allow this ring to dry for 4 min.

(5) While the solution dries, carefully pick 10 worms into a 10-μl drop of M9 on a coverslip. After the ring has dried into the plate, use a micropipet to transfer the drop containing the worms into the center of the ring.

(6) Wait 10 min. At the end of this time, record how many worms are outside this high osmolarity barrier. If you wish, you may observe the worms during the 10 min interval to determine if they back away from the barrier correctly.

(7) Repeat Steps 2–6 with worms from your RNAi plates.

(8) Form a hypothesis to explain the results that you observed in this assay.

Mechanosensation assay (touch-neuron-specific)

This assay is designed to test directly for effects resulting from knocking down genes expressed in the touch neurons. As you may remember from the exercise in Section 1.2, touch neurons respond to mechanical stimulation (gently touching the worm) using an eyebrow hair. Knocking down a gene required for touch neuron function would result in a defective mechanosensation response.

(1) Obtain an eyebrow hair glued to a toothpick from your instructor.

(2) Carefully perform the touch assay as detailed in the exercise in Section 1.2 for 10–15 control animals.

(3) Repeat the touch assay with 10–15 animals from each of your RNAi experiments.

(4) Hypothesize a possible explanation for your observed results based upon your prior knowledge of the target gene.

Muscimol assay (GABAergic-specific)

Muscimol is a chemical agonist of GABA, the most common inhibitory neurotransmitter in most organisms. Muscimol therefore functions to enhance the inhibition normally conveyed by GABA release. In *C. elegans*, the most easily visualized effect of muscimol exposure is that pumping of the pharynx is altered. In worms with a functional GABAergic system, pharyngeal pumping should stop in response to muscimol exposure. If your RNAi experiment has affected GABAergic reception, you should see a change in the rate of response.

(1) Obtain an NGM plate that has been supplemented with 10 mM muscimol from your instructor.

(2) Place 10–15 control animals on the muscimol plate. Allow them to sit for 30–60 min.

(3) Carefully observe the pharynx of each worm using high-power magnification on the dissecting microscope. Record whether or not the pharynx is still pumping. If it is still pumping, count the number of pumps for 1 min.

(4) Repeat with 10–15 animals from each of your RNAi experiments.

(5) Formulate a hypothesis to explain your results based on your knowledge of Gene Y.

8.5.6 Assaying for dauer formation

There are two main dauer formation phenotypes. One class causes worms to enter the dauer stage inappropriately (prematurely) even though environmental conditions may be ideal for normal growth. This can be assayed by a simple visual examination for the formation of dauer animals on a plate with plenty of food that is not overcrowded with animals.

(1) Examine your control plates carefully. Do you observe any dauer larvae? Refer to Figure 1.4 if you need a reminder as to the appearance of dauer larvae.

(2) Repeat this visual examination with each set of your RNAi plates.

(3) Do you observe any dauer animals on any of your plates? Record your observations in your notebook.

(4) Speculate on a possible link between your target gene and dauer larva formation.

The other class of dauer defect causes worms to *fail* to enter the dauer stage even though conditions are no longer ideal for growth. This can be assayed by examining plates with no food for the formation of dauer larvae from recently hatched eggs.

(1) Place 2–5 gravid control animals onto an NGM plate that has not been seeded with food.

(2) Repeat with animals from your RNAi feeding plates; be sure to use a different plate for each set of animals and label them carefully!

(3) Allow the worms to lay eggs for at least 24 h. After that time you may remove the adult animals but it is not necessary.

(4) Wait 3–4 more days and then examine the plates carefully.

(5) Record whether larval worms have entered the dauer stage properly. In other words, are there dauer animals on both your control and RNAi plates?

(6) Form a hypothesis to explain your observed results regarding the formation of dauer larvae.

Instructor's notes

Please note that it is not necessary to perform each assay using worms from each individual plate. Animals on any 'B' plate for a given IPTG concentration are equivalent to animals on the other 'B' plates of that same concentration. You initially set up several plates in order to ensure that you have a sufficient population of animals to work with and in case contamination occurs on some but not all of the plates.

These are several different types of general assays that you may choose to use to analyze your RNAi animals. Although many of the assays above are for examining neuronal defects, this is not meant to insinuate that you will see a large number of neuronal phenotypes. Rather, it is generally thought that neurons are less susceptible to RNAi-induced effects. However, some neuronal defects can be observed. We have outlined numerous neuronal assays simply because these assays are less straightforward than examining for general morphological or developmental defects.

In several of the assays above, students expose worms to different chemicals and test their responses to this exposure. In general, our assays simply test for response to a high level of chemical. However, threshold responses may reveal more subtle defects. Consider titrating the chemicals and testing differences in response along a concentration gradient. Likewise, although wildtype N2 worms are used typically in RNAi, application of this method to specific mutants might yield novel data as well.

The phenotypic possibilities resulting from RNAi in *C. elegans* are both gene-specific and wide-ranging. We encourage students and instructors to investigate the possibilities (in the context of your hypothesis) by looking at Appendix IV for a list of phenotypic defects.

Of course, the instructor or students are strongly encouraged to utilize Wormbase, Wormbook, WormAtlas, and the primary literature to surmise their own best strategies for phenotypic analysis.

References

Fire A, Xu SQ, Montgomery MK, Kostas SA, Driver SE and Mello CC. 1998. *Nature* **391**: 806–811.

Kamath RS, Fraser AG, Dong Y, Poulin G, Durbin R, Gotta M, *et al.* 2003. *Nature* **421**: 231–237.

Piano F, Schetter AJ, Mangone M, Stein L and Kemphues KJ. 2001. *Curr. Biol.* **10**: 1619–1622.

Simmer F, Moorman C, van der Linden AM, Kuijk E, van den Berghe PV, Kamath RS, *et al.* 2003. *PLoS Biol.* **1**: 77–84.

Sonnichsen B, Koski LB, Walsh A, Marschall P, Neumann B, Brehm M, *et al.* 2005. *Nature* **434**: 462–469.

Thomas JH. 1990. *Genetics* **124**: 855–872.

Appendix *I*   Recombinational cloning

Molecular biology was revolutionized by the advent of restriction enzyme cloning in the 1970s, when restriction enzymes were first isolated and shown to cut very specific sequences. Paul Berg of Stanford University first combined the DNA of two different organisms in 1972, ushering in the modern molecular biology age in which now even undergraduates routinely recombine DNA fragments. Despite its many years of use, restriction enzyme cloning can still be an imprecise tool. Several factors must all come together: both the target DNA and the host vector must contain compatible restriction sites; multiple enzymes may need to work together; DNA ligation by the T4 ligase must be efficient; and molar ratios/quantities/sizes of insert and vector DNAs must be favorable for a successful reaction. In all, these factors can result in several weeks for a single DNA cloning reaction to be accomplished. Furthermore, each new cloning reaction of a target DNA into a different host vector must begin anew at the original step of defining compatible restriction maps.

Recombinational cloning was first introduced to molecular biologists in 2000. Patented by the Invitrogen Corporation (www.invitrogen.com), recombinational cloning is often known by the trade name 'Gateway cloning'. In contrast to traditional restriction-based cloning, Gateway cloning is a universal technology that utilizes knowledge from studying the ability of the bacteriophage lambda to specifically recombine pieces of DNA using very specific recognition sequences. The true advantage of the Gateway technology is its versatility. Once the Gateway recognition sequences have been incorporated onto the target DNA, it can then be recombined into any Gateway vector. There is no need to verify

Integrated Genomics   Guy A. Caldwell, Shelli N. Williams, Kim A. Caldwell
© 2006 John Wiley & Sons, Ltd

restriction maps and no need to redesign primers each time a new vector is to be used. The lambda recognition sequences utilized in Gateway cloning are not found naturally, thus preventing the possibility of non-specific recombination. Any type of host vector can be converted to a Gateway compatible vector by the incorporation of the appropriate sequences via traditional cloning methods. After this more time-consuming step has been completed, restriction enzymes are never needed again when cloning into a Gateway-modified vector.

There are two basic Gateway reactions (outlined below): the BP reaction and the LR reaction. Although secondary screening to verify recombination is often unnecessary following a Gateway reaction, a small percentage of false positives (generally less than 8 percent) can result and therefore we recommend that you screen your clones following the LR reaction (and the BP reaction, if time and finances allow). You can screen your clones in one of two ways. You may use the PCR screening protocol detailed in Section 7.6 and then proceed with the mini-prep protocol in Section 7.7. Given the efficiency of recombinational cloning, we recommend going directly to the mini-prop. Alternatively, you may proceed with a check digest as outlined in Section 7.8. If you wish to use the check digest protocol, it will be necessary for you to obtain a restriction map of both the final destination vector and also the target DNA sequence. We recommend choosing an enzyme for your check digest that will cut your target sequence and the vector each one time. You will examine an agarose gel of your digests for the presence of two bands, corresponding to the appropriate sizes based on where the restriction enzyme cut each segment of DNA.

AI.1 Isolation of genomic DNA from *C. elegans*

If using this procedure for generating an RNAi vector, perform the protocol exactly as detailed in Section 7.1 of Chapter 7.

AI.2 PCR amplification of target gene sequence from *C. elegans* genomic DNA

Use the protocol as detailed in Section 7.2 of Experiment 7; however, your primer design will differ from that discussed in the 'Instructor's notes'. Gateway cloning begins with the incorporation of the lambda site-specific *att*achment sites onto the ends of your target DNA sequence. This is performed by simple primer design and subsequent PCR, just as is used to incorporate restriction enzyme recognition sequences for traditional cloning. These serve as the recognition site for the recombination event. Consult the Gateway manual or website (www.invitrogen.com) for incorporating the appropriate sites onto your target DNA sequence.

AI.3 Agarose gel electrophoresis and clean-up of PCR reaction

Follow the protocol in Section 7.3 with the following modifications. Following clean-up of your PCR product using the Qiagen Qiaquick PCR purification kit, you should quantitate your samples. Use the directions outlined in Section 5.4 for quantitating DNA. If possible, proceed immediately from clean-up and quantitation of your PCR product to the procedure in Section AI.4. If you do not proceed immediately, freeze your purified product until you are ready to perform the BP reaction (to clone your target DNA into a Gateway entry vector).

AI.4 Entry vector cloning

After you have amplified the target sequence complete with *attB* sites, reagents from the Gateway cloning kit are used to facilitate the recombination of your sequence into a Gateway donor vector by a BP reaction (*attB* sites are recombined with *attP*). The region originally flanked by *attP* sites within the donor vector is known as the Gateway cassette. Following the recombination reaction catalysed by the enzyme BP clonase, a simple transformation is used to amplify properly recombined vectors. At this time, another advantage of the Gateway technology becomes obvious: vectors that have not undergone recombination will contain the *ccdB* gene, which encodes a protein that interferes with *E. coli* DNA gyrase. The CcdB protein will therefore inhibit the growth of bacteria if it is present within a vector. Only those vectors that have undergone recombination (during which the *ccdB* sequence is replaced with your target sequence) will reproduce following transformation. This greatly reduces the presence of false positives on the transformation plates, thereby decreasing or eliminating the need to screen colonies following transformation of the BP reaction.

Procedural overview

A BP reaction is performed by simply mixing together different components from the Gateway cloning kit with your purified target DNA. During the incubation at 25 °C, the BP clonase enzyme mix facilitates the recombination of your *attB*-flanked target PCR with the pDONR221 vector. This removes the toxic *ccdB* gene from within the vector, allowing properly recombined vectors to be transformed with DH5α cells during the second half of the procedure. These *E. coli* cells have been pretreated to be chemically competent, thereby enabling transformation to occur with high efficiency (see Section 7.5

to review procedural details for a chemical transformation of *E. coli*). The pDONR221 vector carries the gene encoding for resistance to the antibiotic kanamycin, which serves as a second selective marker for transformed cells.

Reagents (see Appendix II for recipes)

Purified target DNA flanked by Gateway *attB* sequences

PCR cloning system with Gateway technology kit (Invitrogen catalog no. 12535-019)

1X TE, pH 8.0

LB agar plate + 50 μg/ml kanamycin

Equipment

Microcentrifuge tubes, 1.5 ml, sterile

Vortex

Ice bucket

25 °C Incubator, stationary

37 °C Incubator (stationary) or waterbath

Snap-cap tubes (such as Falcon 2059), sterile (one per transformation)

42 °C Waterbath

37 °C Incubator, shaking

Bunsen burner or ethanol lamp

Cell spreader

37 °C Incubator, stationary

Day of the lab

NOTE: It is critical to the success of this procedure that reactions and cells be kept on ice as indicated in the protocol.

(1) Obtain your purified PCR product containing the *attB* flanked sequence.

(2) Add the following reagents in the order listed to a microcentrifuge tube on ice:

pDONR221	2 µl
BP reaction buffer	4 µl
PCR product (400–500 ng)	1–10 µl
1X TE, pH 8.0	to 16 µl total volume

(3) Remove BP clonase enzyme from −80 °C and thaw on ice. Vortex the mixture briefly (for 3 s).

(4) Add 4 µl of BP clonase to the above reaction mix on ice. Immediately return the BP clonase to −80 °C.

(5) Vortex the BP reaction mix for 3–5 s. Place at 25 °C in an incubator or waterbath for 1–3 h.

(6) Remove the reaction from 25 °C and add 2 µl of proteinase K to the reaction mix.

(7) Incubate at 37 °C for 10 min.

(8) During the incubation, thaw DH5α cells on ice. Also, place one 15-ml snap-cap tube on ice for each BP reaction.

(9) Aliquot 50 µl of thawed cells into an appropriately labeled snap-cap tube.

(10) Add 1 µl of BP reaction to the tube of cells and incubate on ice for 30 min.

(11) Heat-shock at 42 °C for exactly 30 s. Immediately return tube to ice for 2 min.

(12) Aseptically add 450 µl of room temperature SOC broth to each tube and shake at 225 rpm for 1 h at 37 °C. During this period, label two LB agar plates with kanamycin for each reaction. Label plates on the bottom!

(13) Using a sterile technique and a cell spreader, spread 100 µl of cells from each transformation onto the appropriately labeled plates.

(14) Allow liquid to dry into plates. Incubate the plates, inverted, overnight at 37 °C.

(15) Remove the plates from the incubator after 16–20 h. Store plates at 4 °C until you are ready for the procedure of Section AI.5.

AI.5 Small-scale isolation of plasmid DNA from *E. coli*: the mini-prep

Follow the mini-prep protocol in Section 7.7 of Chapter 7; however, instead of using ampicillin in your overnight cultures, use kanamycin. This is due to the fact that the Gateway entry vector is selectably marked with a gene for kanamycin resistance rather than for ampicillin resistance. This is the only change to the protocol. Also, be sure to quantitate your purified plasmid at the end of the procedure.

AI.6 Destination vector cloning

After you have successfully created and identified a Gateway entry clone (vector pDONR221 containing your recombined target sequence), you are only a few steps away from having a 'ready-to-use' expression vector. At this time, you could perform any number of reactions into any Gateway destination vector. For instance, perhaps you would rather perform a GFP localization study instead of an RNAi feeding experiment, or perhaps you would like to perform both? If this is the case, the same entry clone is suitable for either recombination reaction to generate a plasmid sufficient for such purposes. This inherent flexibility is the prime advantage of recombinational cloning.

Importantly, Gateway-compatible destination vectors for this course are available from Addgene **www.addgene.org/integrated-genius.** *These include L4440, pLeXA, and pACT2.2.*

Procedural overview

The LR recombination reactions works exactly as the BP recombination reaction. The only difference is that the LR clonase enzyme recognizes different attachment sites, in this case *attL* sites, which result from the recombination of the *attB* and *attP* sites during the BP reaction. *Refer to a Gateway cloning manual for figures illustrating the recombination event.* The expression vector also contains a Gateway cassette of the *ccdB* gene flanked by *attR* sites, again reducing false positives by inducing death in cells that have taken up un-recombined vectors. Transformation proceeds as detailed in Section AI.4.

Reagents (see Appendix II for recipes)

Purified entry clone from Section AI.5

LR Clonase enzyme mix (Invitrogen catalog no. 11791-019)

1X TE, pH 8.0

Library efficiency DH5α chemically competent cells (Invitrogen catalog no. 18263-012)

LB agar plate + 100 μg/ml ampicillin

Equipment

Microcentrifuge tubes, 1.5 ml, sterile

Vortex

Ice bucket

25 °C Incubator, stationary

37 °C Waterbath

Snap-cap tubes (such as Falcon 2059), sterile (one per transformation)

42 °C Waterbath

37 °C Incubator, shaking

Bunsen burner or ethanol lamp

Cell spreader

37 °C Incubator, stationary

Day of the lab

NOTE: It is critical to the success of this procedure that reactions and cells be kept on ice as indicated in the protocol.

(1) Obtain your purified entry clone containing the *attL* flanked sequence.

(2) Add the following reagents in the order listed to a microcentrifuge tube on ice:

LR Reaction buffer	4 μl
Entry clone (100–300 ng)	1–10 μl
Destination vector*	1–10 μl
1X TE, pH 8.0	to 16 μl total volume

    *** For this course, destination vector should be a Gateway-converted L4440 vector.**

(3) Remove LR clonase enzyme from −80 °C and thaw on ice. Vortex the mixture briefly (three seconds).

(4) Add 4 μl of LR clonase to the above reaction mix on ice. Immediately return the LR clonase to −80 °C.

(5) Vortex the LR reaction mix for 3–5 s. Place in a 25 °C incubator for 1–3 h.

(6) Remove the reaction from the 25 °C incubator and add 2 μl of proteinase K to the reaction mix.

(7) Incubate at 37 °C for 10 min.

(8) During the incubation in Step 7, thaw DH5α cells on ice. Also, place one 15-ml snap-cap tube on ice for each LR reaction.

(9) Aliquot 50 μl of thawed cells into an appropriately labeled snap-cap tube.

(10) Add 1 μl of LR reaction buffer to the tube of cells and incubate on ice for 30 min.

(11) Heat-shock at 42 °C for exactly 30 s. Immediately return the tube to ice for 2 min.

(12) Aseptically add 450 μl of room temperature of SOC broth to each tube and shake at 225 rpm for 1 h at 37 °C. During this period, label two LB agar plates *with ampicillin* for each reaction. Label plates on the bottom!

(13) Using a sterile technique and a cell spreader, spread 50–100 μl of cells from each transformation onto the appropriately labeled plates.

(14) Allow liquid to dry into the plates. Incubate the plates, inverted, overnight at 37 °C.

(15) Remove the plates from the incubator after 16–20 h to avoid the formation of 'satellite' colonies. Satellites are small colonies resulting from opportunistic bacteria growing within the zone of degraded ampicillin created by true transformants.

Instructor's notes

This protocol could be used for any number or type of recombinational cloning reactions, including the creation of a new yeast two-hybrid bait vector. All that is needed is a vector that has been converted to a Gateway destination vector through the addition of the Gateway cassette. pLexA has been converted for this purpose and is available from Addgene. Please refer to the Gateway manual should you wish to convert any traditional vector to a Gateway vector. Once a vector has been converted to the Gateway system, it is ready for use in this protocol.

AI.7 Small-scale isolation of plasmid DNA from *E. coli*: the mini-prep

Follow the mini-prep protocol in Section 7.7 of Experiment 7. There is no change from the original protocol (you will use ampicillin in the overnight cultures).

As noted above, we recommend secondary screening of your LR clones to be sure that they contain the target DNA sequence. Screening can be performed by PCR screening of colonies or by mini-preps followed by check digest reactions (both protocols are outlined in Experiment 7).

Appendix II Recipes and media preparation

This appendix contains recipes for basic solution preparation as well as detailed instructions for preparing the different types of media used throughout this course. Additionally, there is a protocol within the 'Bacterial media' section for preparing acid-washed receptacles and solutions for making electrocompetent cells.

Solution recipes

Below are recipes for solutions used throughout this course. For a discussion on making basic solutions, consult *Molecular Cloning: A Laboratory Manual (3rd ed)* by Joseph Sambrook and David W. Russell (Cold Spring Harbor Laboratory Press; Cold Spring Harbor, New York, 2001). Solutions are sterilized as detailed at the end of the recipe; if no note about sterilization is present, the solution does not have to be sterilized.

Worm decontamination solution

Typically made in 10 ml quantities. Make fresh before use.

4 percent bleach

$0.5 \, \text{N}$ NaOH

Integrated Genomics   Guy A. Caldwell, Shelli N. Williams, Kim A. Caldwell
© 2006 John Wiley & Sons, Ltd

Freezing solution

For 100 ml:

    1.2 g NaCl

    1.4 g KH_2PO_4

    60.0 g Glycerol

    1.1 ml 1 N NaOH

    Sterilize by autoclaving

    After autoclaving, add 0.6 ml of 0.1 M $MgSO_4$

6 × Gel loading dye

For 10 ml:

    1.5 g Ficoll (Type 400; Pharmacia) in H_2O

    25 mg Bromophenol blue

    25 mg Xylene cyanol FF

    Mix by stirring for 15–20 min

    Store at room temperature in 0.5-ml aliquots

This dye should be used as one-sixth of the final volume of any DNA solution to be electrophoresed. For example, add 2 μl 6X dye to 10 μl DNA solution prior to electrophoresis.

10 × Lithium acetate (LiAc) = 1 M

Sterilize by autoclaving

1 × LiAc–40 percent PEG-3350–1 × TE buffer (typically make 10 ml for 10 transformations)

1 ml 10X LiAc

8 ml 50 percent PEG (mol.wt. 3350)

1 ml 10X TE buffer

Sterilize by autoclaving

M9 buffer

For 500 ml:

    1.5 g KH_2PO_4

    6.4 g Na_2HPO_4

    2.5 g NaCl

    Sterilize by autoclaving

10 × PBS

For 500 ml:

    4.00 g NaCl

    0.10 g KCl

    0.72 g Na_2PO_4

    0.12 g KH_2PO_4 (monobasic)

    pH to 7.4 with HCl

Sterilize by autoclaving

PMB

50 mM Sodium phosphate buffer, pH 7.5

1 mM $MgCl_2$

1 M Potassium phosphate

For 100 ml:

    10.2 g KH_2PO_4 (monobasic) to 75 ml double-distilled water

    5.6 g K_2HPO_4 (dibasic) to 25 ml double-distilled water

    Mix both solutions well, add together and mix well

    Sterilize by autoclaving

25 × TAE buffer stock

For 1 liter:

    121 g Tris base

    10.2 g Sodium acetate

    18.6 g Disodium EDTA

    750 ml Water

    Adjust pH to 7.8 using glacial acetic acid

10 × TE buffer, pH 7.4

100 mM Tris-HCl

10 mM EDTA, pH 7.4

Sterilize by autoclaving

10 × TE buffer, pH 8.0

100 mM Tris-HCl

10 mM EDTA, pH 8.0

Sterilize by autoclaving

Worm β-galactosidase staining solution (prepared fresh before use)

For 10 ml:

    8.4 ml Double-distilled water

    0.5 ml 1 M Sodium phosphate buffer, pH 7.5

    10 μl 1 M $MgCl_2$

    4 μl 10 percent Sodium dodecyl sulfate (SDS)

1 ml 100 mM Stock of potassium ferrocyanide and potassium ferricyanide

62.5 μl 12.5 percent X-gal in $N'N'$-dimethylformamide (DMF)

Worm lysis solution

2.0 percent Triton X-100

1.0 percent SDS

0.1 M NaCl

10 mM Tris-HCl, pH 8.0

1 mM EDTA

26.4 μl Protchase K (19 mg/ml)(cat #1 Roche, 964 372)

2 percent X-gal

200 mg X-gal

10 ml $N'N'$-Dimethylformamide *(wear gloves!)*

Store at −20 °C, wrapped in foil

12.5 percent X-gal

1.25 g X-gal

10 ml $N'N'$-Dimethylformamide *(wear gloves!)*

Store at −20 °C, wrapped in foil

Yeast lysis solution

2.0 percent Triton X-100

1.0 percent SDS

0.1 M NaCl

0.01 M Tris-HCl, pH 8.0

0.001 M EDTA

10 × Z buffer

For 100 ml:

8.5 g Na_2HPO_4

5.5 g $NaH_2PO_4 \bullet H_2O$

0.7 g KCl

1 ml 1 м Mg_2SO_4

Media preparation

Worm media

The instructions below are for making 500 ml of worm media, which will yield 50–60 worm plates that are 60 mm in size rather than the larger 100 mm used for most bacterial and yeast applications. It will be necessary for you to scale up the recipe to have enough plates for the majority of the exercises. However, for volumes above 500 ml it becomes increasingly difficult to pour by hand. Many *C. elegans* labs own a pumping apparatus known as the 'PourBoy' for pouring large volumes of plates. Not only does the PourBoy prevent spillage, it also ensures that the volume of media is consistent from plate to plate. This is very useful when observing many different plates by stereomicroscope (no need for coarse refocusing between plates) and for adding specific concentrations of different chemicals to plates after they have been poured. To order a PourBoy, visit www.TriTechResearch.com. You may also purchase low-cost Petri dishes at this site. It is essential to buy non-vented dishes to prevent desiccation of the media, which occurs rapidly with small volumes. Another option for pouring media is to purchase sterile polypropylene flasks from Corning (catalog no. 431123). You can then transfer small amounts of sterile media to these flasks because they have a convenient opening for pouring sterilized media. These flasks cannot be sterilized and therefore are not reusable.

Making the media

(1) Select a flask of the appropriate size and add water according to the chart below.

(2) Carefully weigh out the ingredients listed in Table AII.1 and add to the water.

(3) Place an aluminum foil cap over the top of the flask.

(4) Mix the ingredients on a stir plate with a magnetic stirbar in the flask; leave the stirbar in the flask (do not remove the stirbar until you are completely finished with the media preparation).

(5) Once the media is mixed well, sterilize by autoclaving.

(6) Following the autoclave cycle, place the flask of media in a 55–60 °C bath for 30 min to cool it evenly. It is important that the media cools to at least 60 °C.

(7) Once the media has cooled, carefully add the ingredients listed in Table AII.2 in the order presented within the table. The media should be stirred while these ingredients are added.

NOTE: $CaCl_2 \cdot 2H_2O$, $MgSO_4 \cdot 7H_2O$ and potassium phosphate stocks should be prepared in advance and sterilized either by autoclaving (potassium phosphate) or by filter-sterilizing. If you choose to filter-sterilize, use a filter with a pore size no larger than $0.2 \mu m$.

Optional: Many labs add the antifungal agent nystatin to worm plates to prevent fungal contamination. If you wish to add this, prepare nystatin according to the information provided in Table AII.3. Once the nystatin is completely in solution, add the entire volume to the appropriate amount of media.

Table AII.1 Preparation of nematode growth media

Media	Flask size	Water	Peptone	NaCl	Agar
100 ml	250 ml	97 ml	0.25 g	0.3 g	1.7 g
300 ml	1 liter	292 ml	0.75 g	0.9 g	5.1 g
500 ml	1 liter	487 ml	1.25 g	1.5 g	8.5 g
1 liter	2 liter	975 ml	2.5 g	3.0 g	17.0 g

Table AII.2 Quantities of supplements required for worm media

Media	Cholesterol (5 mg/ml)	1 M $CaCl_2$	1 M $MgSO_4$	1 M Potassium phosphate	Streptomycin (360 mg/ml)
100 ml	100 µl	100 µl	100 µl	2.5 ml	57 µl
300 ml	300 µl	300 µl	300 µl	7.5 ml	170 µl
500 ml	500 µl	500 µl	500 µl	12.5 ml	283 µl
1 liter	1 ml	1 ml	1 ml	25.0 ml	567 µl

Table AII.3 Nystatin supplementation for worm media

Media	Nystatin powder	100% Ethanol
100 ml	1.1 mg	100 μl
300 ml	3.3 mg	300 μl
500 ml	5.6 mg	500 μl
1 liter	11.2 mg	1 ml

(8) Stir the media for 2–5 min to be sure that all components are in solution.

(9) Carefully pour the media into 60 mm × 15 mm plates using a sterile technique. It is helpful to pour plates in stacks of 5–6 depending on the size of your hand. This cuts down on the amount of space needed for pouring plates and keeps everything well organized.

Preparing plates for C. elegans *growth*

(1) Allow the media to solidify for at least 24 h without moving the plates. This will ensure that the media forms a level surface. After 24 h it will be necessary to remove condensation from the lids of the top few plates in each stack. Condensation forms on the lids as the plates cool. Specifically the plates at the top of the stack, which are exposed to the cooler air, form condensation.

(2) To remove condensation from the lids of plates, carefully pick up the plate without knocking the condensation onto the media. Hold the base of the plate (containing the media) in one hand and grasp the lid with the other hand. In one motion, lift the lid and knock it against a benchtop to shake off the condensation. At the same time, tilt the bottom of the plate down so that the bottom of the Petri dish is facing upwards to the air (media surface is facing down). This will prevent airborne contaminants from falling down onto the growth surface of the worm media.

(3) At this point, you may store the plates in an airtight container (Tupperware bins work very well) at 4 °C for several months. To store plates, carefully invert the plates (media side up) within the bin. This prevents excess condensation from forming on the lids and resulting in dry, thin plates. Alternatively, you may 'seed' the plates (as below) with a bacterial lawn that worms will eat and either use immediately or store them, seeded, at 4 °C.

Seeding plates

(1) You may seed plates within 1–3 days after removing the condensation from the plates. If you do not plan on seeding within 3 days, it is recommended that you store the plates as detailed above in an airtight container.

(2) Before seeding plates you will need a saturated culture of the *E. coli* strain OP50. This strain can be obtained from the CGC at http://biosci.umn.edu/CGC/ CGChomepage.htm. The recipe above requires that you have obtained the OP50 strain carrying a streptomycin-resistance gene. It is highly recommended that you use this strain so that you can include the antibiotic in the media, because new students frequently contaminate plates.

(3) To prepare a saturated liquid culture of OP50 for seeding worm plates, inoculate 50 ml of B broth (see section below on 'Bacterial media') supplemented with streptomycin by aseptically adding bacterial growth from an existing worm plate to the liquid using a sterile bacteria loop. Use a fresh worm plate to prevent cross-contamination. Alternatively, you could obtain a starter culture from a frozen glycerol stock of OP50. Grow overnight at 37 °C. The culture is ready for use the following morning but may be stored for several weeks at 4 °C. It is best to inoculate the bacteria in a sterile bottle of at least 100-ml volume.

(4) When seeding worm plates to generate a bacterial lawn, it is important that the bacteria not be allowed to touch the walls of the dish. If the bacterial lawn touches the walls, the worms will likely crawl through the bacteria, up the walls and out of the plate. To seed plates that will be used for mating, simply use a smaller bacterial lawn. The smaller lawn will keep the worms in closer proximity and increase the chances of a successful mating.

(5) To seed worm plates, place a pipet bulb on the end of a sterile 5-ml glass pipet. Carefully draw up a pipet full of liquid OP50. Being careful not to scratch the surface of the agar, lightly spread bacteria over the surface of the media. Do not allow the bacteria to touch the edge of the plate and do not spread the bacteria too thickly. Allow the bacteria to soak into the plate and then incubate overnight at 37 °C. Do not invert the plates.

(6) After seeding plates and growing overnight, they may be stored inverted in an airtight container (Tupperware is ideal) for several months before use (see Figure AII.1 for a representative image of a nicely seeded worm plate).

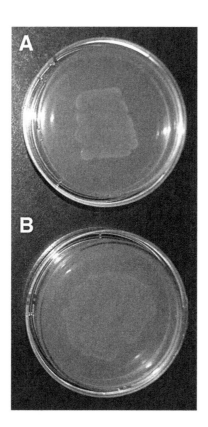

Figure AII.1 Examples of well-seeded 60-mm worm plates. Note that both have bacterial lawns located in the center of the plate: (A) a small bacterial lawn for mating experiments; (B) a normal sized lawn for standard worm maintenance

Supplementing NGM plates with chemicals

When making plates to which you will later add drugs, the most important point to note is that you must know precisely how much media is present within each plate. If you use a PourBoy to make plates, simply be sure that you calibrate your setting before beginning. If you do not use a PourBoy, proceed using the directions below.

(1) Make media as directed above through Step 8 of the 'Making media' section. It is recommended that you make only 100 ml of media at a time if you do not use a PourBoy.

(2) After you have added all extra ingredients to the media solution, obtain a sterile 10-ml glass pipet.

(3) Carefully aliquot 10 ml of media into ten 60 mm × 15 mm plates using a sterile technique. You will need to work quickly so that the agar does not solidify within

the glass pipet. If the agar does begin to solidify in the pipet, quickly place the pipet into a container of hot water to melt the agar. You will need to use a different, sterile pipet for subsequent dispensing of any remaining media.

(4) Allow the media to dry for 24–48 h. At this point, you can store the plates in an airtight container (Tupperware bins work very well) at 4 °C for several months. To store plates, carefully invert the plates (media side up) within the bin. This prevents excess condensation from forming on the lids and resulting in dry, thin plates.

(5) To supplement plates with chemicals you will need to make a liquid stock of the chemical to be added to the plate. Listed below are two chemicals used in the RNAi assays outlined in Section 8.5 as well as how much of the drug to add to a single 10-ml NGM plate.

Chemical	Stock solution	Add
Muscimol	1 M	100 μl
Aldicarb	500 mM	10 μl

Yeast media

There are several points to note when pouring yeast plates. First, because yeast plates do not contain antibiotics, they are easily contaminated. Before pouring yeast plates, carefully wipe the working surface down with 70 percent ethanol. As mentioned below, a tissue culture or UV-irradiated hood is preferable for pouring plates. Secondly, yeast grow slower than bacteria, therefore the plates will remain in incubation for long times and should be poured thicker than bacterial plates (∼30–35 ml per plate). It should be noted that the most common medium for growing *S. cerevisiae* is called 'SC' or Synthetic complete medium.

SC plates

The recipe below makes *1 liter* of agar (∼36–38 plates) and can be used for the growth of *S. cerevisiae* strain L40.

SC media in broth or agar plates

As discussed in the introduction to Experiment 3 (yeast two-hybrid system), growth deficiencies are used as selectable markers in yeast strains just as antibiotic resistance is used in bacterial strains. Before preparing this media, be sure that you know what

Table AII.4 Preparation and storage of common amino acid stock solutions

Amino acid	Storage	Sterilize by
1.0% Tryptophan	In foil at 4 °C	Filter
1.0% Leucine	Room temp.	Autoclave
1.0% Histidine	4 °C	Autoclave
0.2% Uracil	Room temp.	Autoclave
1.0% Lysine	4 °C	Autoclave

type of selective media you will need (i.e. which amino acids you must provide for the yeast and which should be left out as a selective pressure). Additionally, you will need to prepare amino acid stocks according to the information provided within Table AII.4. Typically 100-ml stock solutions are made.

Different amino acid powder mixes (or 'drop-out' media) are available for the growth of different yeast strains. One can supplant specific amino acids by adding back certain ones from stock solutions. For example, SC-Trp media (which would correspond to SC complete media lacking only tryptophan) would be made using the SC-Trp-Leu-His-Ura-Lys amino acid powder, but then Leu, His, Ura, and Lys can all be supplanted by subsequent addition of the appropriate quantity of amino acid solution (as listed below from the separate stocks of those specific supplements). Alternatively, one could purchase a drop-out mix that is only SC-Trp, alleviating the need to add any amino acid stocks to the media. This allows the investigator flexibility in preparing media for defined growth selection and maintenance of specific plasmids within yeast cells.

(1) Weigh out 17.0 g of agar and dissolve in 500 ml of double-distilled water in a 2-liter flask. Leave the stirbar in the flask.

(2) Autoclave for 20 min.

(3) Equilibrate in a 55–60 °C bath for 20–30 min.

(4) While the agar is equilibrating, combine the following in a 1-liter beaker:

20.0 g Glucose

6.7 g Yeast Nitrogen Base without amino acids* (Sigma # 0626)

1.4 g SC-Trp-Leu-His-Ura

500.0 ml double-dsitilled water

Apply low to medium heat if necessary to mix completely into solution.

*If you use a different drop-out mix, the amount added may differ. In all cases, follow the directions listed on the bottle.

(5) Mix well and *add the following* quantity of stock amino acid solutions as necessary for the selection desired:

> 2.0 ml of 1.0 percent tryptophan
> 10.0 ml of 1.0 percent leucine
> 2.0 ml of 1.0 percent histidine
> 10.0 ml of 0.2 percent uracil
> 3.0 ml of 1.0 percent lysine

(6) *Filter-sterilize* using a 500-ml 0.2 μm filter unit (or two 250-ml units).

(7) Place filtered solution in a 55–60 °C bath for 10–15 min to equilibrate with the agar.

(8) Using sterile technique (see Appendix II), combine with agar into the 2-liter flask and stir gently to avoid forming bubbles. Pour ~36–38 100-mm plates (in a tissue culture hood, if available) relatively thick because yeast require time to grow and plates will crack if poured too thin.

(9) Allow plates to dry for at least 24 h but no more than 48 h.

(10) Label your plates appropriately. It is often useful to generate a 'code' for different varieties of your plates. For example, one red and one blue stripe on the side of the plastic lid/bottom could indicate SC-Trp-Leu. Different combinations of laboratory marker colors could then be used for different amino acid deficiencies.

(11) At this point, you may either store than wrap plates in parafilm *or* in an airtight container (Tupperware bins work very well) at 4 °C for several months. To store plates, carefully invert the plates (media side up) within the bin. This prevents excess condensation from forming on the lids and resulting in dry, thin plates.

SC broth

To make broth, use the recipe detailed above with twice the double-distilled water *but no agar.*

Instructor's note

While *S. cerevisae* L40 grows on SC median, it is *not* the optimal growth medium noted for this strain. HSM (Hollenberg Supplement Mixture) has been noted as ideal for this specific strain. However, commercial suppliers have discontinued offering drop-out mixes for HSM at the time of press for this text. Thus, the instructors are encouraged to consider making their own mix if they notice problems with L40 growth or transformation efficiencies on SC plates/broth. Recipies for HSM are available in this reference: Vojtck, A.B. and Hollenberg, S.M. "Ras-Raf interaction two-hybrid analysis" *Methods Enzymol.* 1995; 255:331–92.

YEPD broth

For 1 liter:

   10 g Yeast extract

   20 g Peptone

   20 g Dextrose

   Sterilize by autoclaving

Bacterial media

B broth

This media is used for growth of bacteria for worm maintenance. The recipe makes 500 ml, which should be aliquoted into 50-ml volumes in 100-ml bottles. It is important to know exactly how much B broth is in each bottle because you will need to add streptomycin to the media prior to inoculating it with OP50.

(1) Combine the following ingredients into a total volume of 500 ml of double-distilled water in a beaker:

   5 g of tryptone
   2.5 g of NaCl

(2) Once the ingredients are completely in solution, aliquot into individual bottles (suggested volume of 50 ml) and sterilize by autoclaving.

LB broth

See below for a recipe for making LB agar plates. To make broth, use this recipe with twice the double-distilled water but no agar.

LB + antibiotic plates

This recipe makes 1 liter (~40–50 plates at 20–25 ml per plate). The media is used for general bacterial procedures such as recombinant DNA cloning.

(1) Combine the following ingredients into a total volume of 1000 ml of double-distilled water in a 2-liter flask:

> 10 g of tryptone
> 5 g of yeast extract
> 10 g of NaCl

(2) Use 5 N NaOH to bring the pH of the media to 7.0.

(3) Add 15 g of agar to the flask and cover with an aluminum foil cap.

(4) Autoclave for 20 min.

(5) After the autoclave cycle, cool the media in a 55–60 °C bath for 20–30 min.

(6) When the media has cooled sufficiently (60 °C or below, so that you can comfortably hold the flask), carefully add appropriate amounts of filter-sterilized antibiotic from stock solutions as needed for your application. See below for commonly used concentrations.

(7) Using a sterile technique, carefully pour the media into 100 mm × 15 mm plates.

(8) Allow plates to dry for at least 24 h but no more than 48 h.

(9) At this point, you may store the plates in an airtight container (Tupperware bins work very well) at 4 °C for several months. To store plates, carefully invert the plates (media side up) within the bin. This prevents excess condensation from forming on the lids and resulting in dry, thin plates.

These plates can be poured thinner than yeast plates (~25 ml of media per plate) because bacteria are typically not grown for extended periods of time, as yeast are.

Common antibiotic concentrations and stocks are listed in Table AII.5. It is best to make 10–50 ml of an antibiotic at a time and aliquot into 1-ml aliquots in sterile 1.5-ml

Table AII.5 Preparation and storage of common antibiotic stock solutions

Antibiotic	Dissolved in	Stock	Working concentration
Ampicillin	Water	100 mg/ml	100 μg/ml
Kanamycin	Water	10 mg/ml	50 μg/ml
Tetracycline*	Ethanol	5 mg/ml	12.5 μg/ml
Streptomycin	Water	360 mg/ml	204 μg/ml

*Note that tetracycline is light sensitive, so stocks and plates should be kept in the dark.

microcentrifuge tubes. This prevents degradation of the antibiotic following multiple freeze–thaw cycles.

M9-Leu + Amp agar

This recipe makes 1 liter (~ 40 plates) and is used for recovery of the library plasmid from the yeast two-hybrid screen.

(1) Add 15 g of agar to 775 ml of double-distilled water, cover with an aluminum foil cap and autoclave for 20 min.

(2) After the autoclave cycle, cool the media in a 55–60 °C bath for 30–60 min.

(3) Prepare 1 × M9 Minimal Salts (Difco catalog no. 0485-17): add 2.26 g of M9 Minimal Salts powder to 200 ml of double-distilled water and filter-sterilize. Place in a 55–60 °C bath for 10–15 min to equilibrate with the agar.

(4) Combine the following in a 50-ml Falcon tube:

   20 ml of 20 percent glucose (sterilized stock)
   100 μl of 1 M $CaCl_2$ (from filter-sterilized stock)
   2 ml of 1 M $MgSO_4$ (from filter-sterilized stock)
   500 μl of ampicillin (from 100 mg/ml stock)

(5) Combine the solutions from Steps 3 and 4 with the agar from Step 2. Mix well and pour using an appropriate sterile technique into 100 mm × 15 mm plates.

(6) Allow plates to dry for at least 24 h but no more than 48 h.

(7) At this point, you may store the plates in an airtight container (Tupperware bins work very well) at 4 °C for several months. To store, carefully invert the plates (media side up) within the bin. This prevents excess condensation from forming on the lids and resulting in dry, thin plates.

M9-Leu + Amp broth

Mix the following to make 200 ml solution:

195 ml double-distilled water

452 mg M9 Minimal Salts (Difco # 0485-17)

4 ml 20% glucose (from sterilized stock)

20 μl 1M $CaCl_2$ (from filter-sterilized stock)

400 μl 1M M_gSO_4 (from filter-sterilized stock)

100 μl Ampicillin (100 mg/ml stock)

Filter sterilize, Store at 4 °C.

SOC medium

For 1 liter:

0.5 g NaCl

5 g $MgSO_4 \cdot 7H_2O$

20 g Tryptone

5 g Yeast extract

20 mM Glucose

Sterilize by autoclaving

Acid washing

Acid washing is necessary when preparing electrocompetent bacterial cells. By acid washing all receptacles used for making solutions and for holding cells, you are removing extraneous ions from the bacterial environment. This is important because excess charged molecules within the final bacterial preparation can lead to arcing during electroporation. Arcing will result in an unsuccessful transformation because it kills the cells.

(1) Gather the following together:

   50-ml glass flask
   2-liter glass flask (2)
   1-Liter glass flask*
   250-ml Polypropylene centrifuge bottles (two)

50-ml Polypropylene centrifuge tubes (two)

100-ml Glass bottles for storing double-distilled water and 10 percent glycerol (two)

Autoclavable plastic rack that normally contains tips but without any tips in it

Autoclavable small plastic beaker that will fit over the top of a 2-liter glass flask

100-ml Graduated cylinder

1-Liter graduated cylinder

You should not use an aluminum foil cap on the flask because it will introduce charged ions into the culture. Instead use autoclavable plastic caps such as plastic beakers for covers.

(2) Make 6 liters of 0.25 M HCl. Remember, always add acid to water, never water to acid.

(3) Wearing gloves, carefully fill all items from Step 1 with 0.25 M HCl. Place the lids to the bottles and tubes into the plastic rack that is also filled with HCl.

(4) Let these item to sit for 30 min.

(5) Rinse everything four times with distilled water.

(6) Rinse everything four times with deionized water.

(7) Fill everything with deionized water and autoclave for 25 min on slow exhaust. Do not cap the plastic bottles or tubes but you can loosely put the lids on the glass bottles and invert the plastic beaker onto the opening of the 2-liter flask.

(8) When the autoclave cycle is complete, pour out all the water. Autoclave again for 25 min on fast exhaust.

(9) When this autoclave finishes, carefully place the lids on the plastic bottle and tubes but do not tighten all the way.

Preparing media and solutions for electroporation

NOTE: Do not use metal spatulas or scoops for weighing out reagents or stirbars at this point. They may introduce ions into the media, which could result in arcing when the cells are used for electroporation.

(1) Make 1 liter of HEPES buffer. Add 0.283 g of HEPES to 1 liter of water in the 1-liter flask. Measure the water with the acid-washed graduated cylinder. Swirl the beaker by hand until the HEPES is in solution. Filter-sterilize the solution into two 500-ml Millipore filter units with a pore size of 0.2 μm.

(2) Make 1 liter of LB broth. Add the following ingredients to the 2-liter flask:

> 10 g of tryptone
> 5 g of yeast extract
> 10 g of NaCl
> 950 ml of double-distilled water

Swirl the flask by hand until the ingredients are in solution. Bring the volume up to a total of 1000 ml using the acid-washed 1-liter graduated cyclinder. Place the plastic beaker over the lid of the flask.

(3) Place 50 ml of double-distilled water into one acid-washed 100-ml bottle.

(4) Make 50 ml of 10 percent glycerol and place it in the other acid-washed 100-ml bottle.

(5) Autoclave LB, double-distilled water, and glycerol for 45 min on slow exhaust.

(6) Once the solutions have cooled, they are ready to use for preparing electrocompetent bacterial cells (see Section 5.1).

Appendix III
Sterile techniques and worm protocols

Sterile techniques

Sterile techniques are of the utmost importance in the laboratory setting. These microbiological practices prevent the spread of potentially biohazardous strains or compounds to the scientist and to the environment. They also prevent the introduction of foreign organisms into the experimental setting, which may have adverse effects on the outcome of a study. The basis of aseptic techniques is the rule that no direct contact should be made with the experimental organisms or cultures. In other words, no contact is made with the skin, nose, eyes, mouth or even the work bench to the extent possible. One should always work under the general assumption that any organism or culture media is potentially dangerous. Even if this is not true (i.e. *C. elegans* is relatively harmless), utilizing aseptic techniques should become second nature within any laboratory environment that employs microorganisms as research organisms.

Below are rules to follow during all manipulations of microorganisms. In some cases, such as the transfer of worms, it is not possible to work with a completely sterile technique, hence worm transfer should be treated as a semi-sterile exercise in which you work as quickly as possible to minimize exposure to the environment.

Integrated Genomics   Guy A. Caldwell, Shelli N. Williams, Kim A. Caldwell
© 2006 John Wiley & Sons, Ltd

General considerations for working with microorganisms

(1) There is to be no eating, drinking, chewing gum, or smoking in the laboratory.

(2) Avoid touching the face, eyes, and other exposed areas of the body.

(3) There should be no chewing or biting of pencils and pens.

(4) Keep the work area of the bench clear, clean, tidy, and free of non-laboratory and unessential items.

(5) Nothing should be removed from the laboratory without the direct permission of the laboratory coordinator or safety official.

(6) Contaminated glassware should be placed in lidded receptacles when possible.

(7) Discarded Petri dishes should always be placed in a waste container that can be sealed.

(8) Realize that certain procedures or equipment (agitation of fluids in flasks, etc.) produce aerosols of contaminated materials. Keep lids on all culture vessels when possible.

(9) Any accident, no matter how insignificant (scraped finger or spilled culture fluid), must be reported to the laboratory coordinator or safety official.

(10) Always clean up your work area before leaving the lab. This includes placing all waste in the proper disposal area and returning all supplies to their normal storage area. You should also clean the surface of your workbench with a disinfectant.

(11) Whenever you leave the laboratory, wash your hands with a germicidal soap and dry them with disposable towels.

Guidelines for specific manipulations of microorganisms

(1) When working with inoculating loops, pipets, or agar plates never touch the loop, the tip of the pipet, or the surface of the plate, nor should any of these be placed onto contaminating surfaces (which are any surfaces exposed to the general environment).

(2) Reduce the possible production of hazardous aerosols from cultures as much as possible. When manipulating with a pipet or inoculating loop, keep material away from face.

(3) Never mouth-pipet any liquid within a laboratory. Pipet bulbs or automatic pipets should be used to transfer liquid when necessary.

(4) Always use sterile equipment when manipulating cultures. All pipets should be sterile and aseptic techniques should be utilized during manipulation to preserve this state. After transferring cultures, pipets should be immersed in a suitable disinfectant.

(5) Whenever culture liquid is to be poured from a container, the lips of the tubes and flasks should be flamed with an ethanol lamp or Bunsen burner. This creates warm air convection currents, which will help to prevent contaminating substances from falling into the open mouth of the container.

(6) The lids of Petri dishes should always be placed down on benchtops when they must be removed. Do not turn the lid up so that the inner surface is exposed to the air above.

(7) Whenever a culture container must be opened, hold the container at a 45° angle so that nothing can fall directly into the mouth of the container.

Worm protocols

The protocols outlined below are several of general use within any lab utilizing *C. elegans* as a study organism. Although these protocols are not scheduled directly in this course, they will prove invaluable to the set-up and long-term maintenance of the strains used each time this course is taught. In particular, the protocol outlining freezing of worm strains should be used at the end of the course each time it is taught. By preserving these worm lines from year to year, you will not need to contact and reorder strains from the Caenorhabditis Genetics Consortium each time the course is offered.

How to make a worm pick

Worm picks are made of a thin piece of platinum wire attached to the end of a shortened glass pipet. Platinum wire is used because it can be heated rapidly and cooled for sterilizing the end that will touch the worm and is sturdy enough to withstand numerous heating/cooling cycles. Platinum wire with a diameter of 0.20 mm and AWG Gauge 28–30 can be ordered from Umetals.com (cat. # 14242) or Tritech Research (cat. # PT-9010-010) the price will vary based upon the current market price of platinum. Each worm pick requires approximately 2–3 cm of platinum.

Supplies

6-inch glass Pasteur pipet

Platinum wire, cut into pieces 35–40 mm long

Forceps with rubber grips (Millipore catalog no. XX6200006)

Bunsen burner or ethanol lamp

Recommended

Glass file

Safety goggles

(1) Using either a sharp pressure with your fingers or a glass file, break a glass pipette approximately 10 mm from the point where the pipet begins to contract. Discard the broken off tip. If you choose to break the glass with your fingers, be sure to wrap a paper towel around the pipette to avoid injury.

(2) Hold the broken end of the pipette within a flame, turning it every few seconds to distribute the heat evenly.

(3) Once the end begins to contract inwards, forming a smaller bottleneck, carefully insert the platinum wire into the hole using forceps. Insert only the bottom 1/3 of the wire into the glass.

(4) Continue heating the glass and wire evenly. Soon the wire should stick to the glass as it heats and you will no longer need to hold it in place with the forceps.

(5) Use the forceps to squeeze the end of the glass pipette together around the platinum wire. Continue to do this, while turning the pipette and squeezing gently from all directions until you have completely sealed the hole around the wire.

(6) Allow the glass to cool completely. Gently tug on the wire to ensure it is firmly enclosed within the glass.

(7) Once you are sure that the wire is firmly attached, use a small coin to flatten the end of the wire that will be used to transfer worms. Press the wire down on a flat surface using the coin and raise the other end of the pick so that you create a flat platform that will be parallel to the surface of a worm plate when you are holding the glass pick in your hand. Think of the very end of the worm pick as a spatula and attempt to mold the wire to create this shape. Once you have created the platform, use the coin to widen the transfer surface by pressing side-to-side until the end is

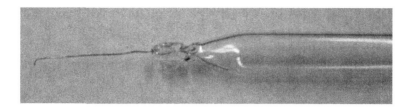

Figure AIII.1 Photograph depicting a worm pick

wider and thinner than the rest of the wire, See Figure AIII. 1 for a representative image of a nice worm pick.

Decontaminating worm stocks

Within a busy laboratory environment, worm stocks often become contaminated with extraneous bacteria, mold, or fungus. It is possible that contamination can have an adverse effect on the worm stock or experimental scheme, therefore at the first sign of contamination it is recommended that you decontaminate your stocks.

Furthermore, RNAi feeding is based upon the premise of ingestion of a specific bacterial strain for delivery of dsRNA. The slightest sign of contamination requires that you discard those plates and animals because you cannot be sure that worms have eaten dsRNA-producing bacteria. In order to prevent transfer of contamination from stock plates, we advise that you frequently decontaminate the stocks of animals that you will use for RNAi feeding. If you will be performing only an occasional round of RNAi feeding, decontaminate stocks 2 weeks before you will be using those animals. If you perform frequent, multiple rounds of RNAi feeding, we recommend that you decontaminate your stocks every 1–2 weeks.

(1) Strip 2–4 fresh NGM plates from dauer animals of the strain that you wish to decontaminate. Be sure that you have labeled the new plates carefully.

(2) Monitor the worms carefully; you will need gravid adults for decontaminating stocks.

(3) Once the animals have developed into gravid adults (you should be able to see dark embryos within the slightly distended body of the animal), carefully transfer 5–10 animals onto a fresh plate, labeled appropriately. Work quickly or pick up several worms at once so that the animals are grouped together on the new plate.

(4) Immediately place 5 μl of decontamination solution directly onto the gravid animals on the fresh plate.

(5) Repeat Steps 3 and 4 for several new plates or for several additional groups of animals onto the same plate.

(6) Allow the decontamination solution to soak into the plate before you invert the plate for normal growth.

Maintaining worm stocks

Although the dauer stage is an invaluable one for maintaining worm stocks, it is not an indefinite storage stage. Eventually plates will begin to dry out and animals will crawl away or even die, therefore in order to maintain your animals in a reasonable working state we recommend that you perform basic maintenance on your stocks every 4–5 weeks.

(1) Obtain a dauer plate representing each stock of animals.

(2) Carefully strip onto 1–2 fresh plates for each strain. Be sure that you have labeled each new plate appropriately.

(3) After stripping each stock, carefully wrap the edge of each plate with a thin strip of parafilm. You should cover the space left between the lid and bottom of the plate with the parafilm.

(4) Place the plates at the appropriate growth temperature; if your worm strain does not require a set growth temperature, 15 °C works well for long-term maintenance of animals.

(5) One to two weeks before you will need a strain, we recommend that you strip several fresh plates to ensure that the worms are healthy, growing well, and free of contamination.

Freezing worm stocks

In addition to periodically stripping worms as a means of stock maintenance, an additional option exists for *C. elegans* researchers. Much like bacteria and yeast, *C. elegans* can be frozen for long-term storage at −80 °C. However, only worms frozen as L1 larvae will survive the freezing process. The general idea for freezing worm strains is to freeze slowly and thaw rapidly. The freezing process is slowed by the use of an insulating layer surrounding the tubes. Rapid thawing is facilitated by heating the tube using body temperature. Use the following protocol for setting up plates for freezing stocks, and the lifecycle chart (Figure 1.1) for timing.

(1) Carefully strip 3–5 fresh plates of animals from dauer plates of each stock. Be sure to use well-labeled plates.

(2) Monitor the animals carefully; when they are young adults, carefully pick 10–15 animals onto a fresh plate labeled appropriately.

(3) Repeat with three more plates for a total of four.

(4) Monitor the plates carefully. The worms should begin laying eggs within hours. Eggs will hatch at different rates. When the majority of the young are L1 or unhatched eggs, the plates are ready for freezing.

(5) Carefully wash each plate with M9 using a glass Pasteur pipet. Use 1.5-ml total M9 volume for washing all four plates. Combine the liquid from the washes into a single 15-ml Falcon tube.

(6) Carefully add an equal volume of freezing media (1.5 ml) for a total of 3 ml of liquid.

(7) Flick the tube with a fingertip to mix the liquid and worms.

(8) Use a glass Pasteur pipet to aliquot equal volumes of the worms in freezing media into five cryogenic tubes, labeled carefully with strain information.

(9) Place the five tubes into a Styrofoam 'sandwich'. The stand that the 15-ml Falcon tubes come in works well for this. Place the tubes into one Styrofoam stand and then enclose them within a second stand inverted and placed over the top. Use a rubber band to keep the 'sandwich' closed, so that the tubes are enclosed by Styrofoam on all sides.

(10) Place at $-80\,°C$.

(11) Two to three days later, remove one tube from $-80\,°C$. Hold it in your hand to rapidly thaw the worms. When the liquid has completely thawed, dump it out onto a new plate, labeled appropriately.

(12) One to two days later, check the plate containing the thawed animals. Be sure that you can see revived animals crawling around on the plate. If so, your freezing was successful and the remaining four tubes can be placed in a box for long-term storage at $-80\,°C$.

Temperature-dependent Growth of C. elegans

As discussed in Section 1.1, the *C. elegans* growth rate is dependent upon the temperature at which the animals are grown. During the course of your studies, should you need to decrease or increase the rate at which your animals are developing, refer to the lifecycle at 25 °C (Figure 1.1) or the lifecycles at 20 °C or 15 °C depicted in Figures AIII.2 and AIII.3, respectively.

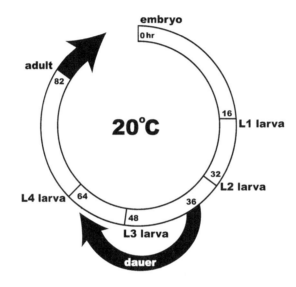

Figure AIII.2   *Caenorhabditis elegans* lifecycle at 20°C

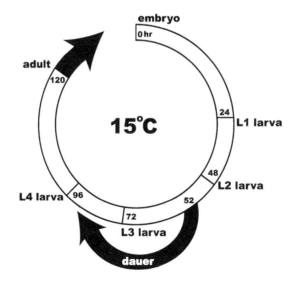

Figure AIII.3   *Caenorhabditis elegans* lifecycle at 15°C

Appendix IV — Mutant C. elegans phenotypes

Below is a list of possible phenotypes that you may be able to observe in select *C. elegans* mutants and following RNA-mediated interference (RNAi) of your target Gene Y. The phenotypes are grouped roughly into associated categories (behavioral, morphology, etc). Your instructor may provide assays for examining these different classes of defects. Phenotypes marked with an asterisk(*) are only visible under high-power magnification using a compound microscope equipped with DIC optics.

Behavioral phenotypes

Adp (sensory **ad**a**p**tation abnormal): Worms display a diminished response to odorants.

Aho (**a**bnormal **h**unger **o**rientation): Worms can associate feeding state (starved versus plentiful food source) with the temperature at which they are grown. Worms displaying Aho phenotypes are unable to associate properly the cultivation temperature with feeding state.

Bor (**bor**dering behavior): N2 animals are normally solitary worms when well fed but occasionally may be seen together at the edges of bacterial lawns. Animals displaying a bordering behavior will accumulate almost exclusively at the edge of the food lawn and often aggregate together in clumps of 10–50 worms or even more.

Integrated Genomics Guy A. Caldwell, Shelli N. Williams, Kim A. Caldwell
© 2006 John Wiley & Sons, Ltd

Che (**che**motaxis abnormal): Worms are normally attracted to bacteria and to various chemicals through receipt of chemotaxic signals via neurons located in the head of the animal. Failure to respond to these attractants is classified as a Che phenotype.

Dec (**de**fecation **c**ycle abnormal): Dec animals display an altered rhythm of defecation. Phenotypes are further divided into Dec-s and Dec-l phenotypes. Dec-s indicates a shortened cycle of defecation whereas Dec-l will manifest as a prolonged rhythm of defecation. An assay for this phenotype is described in detail in Section 8.5.4.

Eat (**eat**ing abnormal): Animals with Eat phenotypes show defects in pharyngeal pumping; this is visible under a dissecting microscope. These defects may include slower or faster pumping rates. Typically a wildtype worm has between 150 and 300 pumps per minute, depending on the exact age of the animal.

Egl (**eg**g-**l**aying abnormal): Egl animals retain eggs within the body cavity of the animal rather than laying them in a wildtype fashion. Egl animals often appear swollen due to the presence of accumulated eggs within the uterus.

Exp (**exp**ulsion defective): Mutants in this class of phenotypes display defects in the defecation cycle, specifically in expulsion of waste. Exp animals are often constipated as a result of defects in this cycle; their intestines will appear distended as a result.

Fab (**f**oraging behavior **ab**normal): Fab animals forage hyperactively for food, disrupting the normal wildtype rhythm of foraging.

Hab (**hab**ituation abnormal): *C. elegans* will adapt to repeated mechanical stimuli such as a touch or tap on the head. Hab mutants display abnormal responses to repeated stimuli; these responses may be slower, more rapid, or normal but incomplete habituation.

Hen (**he**sitatio**n** behavior): Worms displaying a Hen phenotype will hesitate in traveling toward an attractant chemical when encountering an aversion barrier of an ionic solution.

Itt (**i**ncreased **t**hermo**t**olerance): Worms are able to survive at much higher temperatures (i.e. 35 °C) if they display an Itt phenotype.

Lev (**lev**amisole resistance abnormal): Lev animals are resistant to the normal paralysis caused by levamisole exposure. Levamisole is a cholingeric antagonist, thus Lev mutations are generally but not always caused by mutations in acetylcholine-related genes.

Mec (**mec**hanosensory abnormality): Worms display a defective response to touch stimuli using an eyebrow hair. An assay for this phenotype is described in Section 1.2.

Mod (**m**odulation **o**f locomotion **d**efective): When encountering a bacterial lawn, wildtype animals slow their forward locomotion; this slowing is enhanced following 30 min of food deprivation (worms slow even more). Mod animals do not slow in the presence of a lawn of food following 30 min of deprivation; this movement is regulated by the serotonin pathway.

Not (**no**se **t**ouch response defective): Worms fail to respond to a light touch on the tip of the nose. This is different from a Mec phenotype in which touch at the anterior and posterior region fails to generate movement in either direction.

Odr (**od**o**r**ant response abnormality): Odr mutants fail to chemotax to volatile compounds. The normal response to a particular compound could be either an attraction or repulsion.

Osm (**osm**otic avoidance abnormality): Worms with Osm phenotypes fail to avoid regions of high osmotic strength such as 4 M NaCl. This phenotype is described in Section 8.5.5.

Swa (**sw**imming **a**bnormal): Worms are unable to initiate or maintain movement in liquid.

Tab (**t**ouch **ab**normal): Worms with Tab phenotypes exhibit more severe mechanosensation defects and do not respond to a touch by either an eyebrow hair or a thicker wire pick; in contrast, Mec worms will respond to a touch by a wire pick.

Tax (chemo**tax**is abnormal): Worms fail to chemotax toward any attractant molecules.

Ttx (**t**hermo**t**a**x**is abnormal): Wildtype worms will gravitate toward their temperature of cultivation when exposed to a thermogradient. Ttx animals do not migrate to their cultivation temperature.

Unc (**unc**oordinated): Unc animals do not move in the normal sinusoidal pattern of wildtype animals. An assay for this phenotype is described in Section 8.5.4.

Temporal or lifespan phenotypes

Age (**age**ing alteration): Worms display defective lifespan; these may represent either an extended or shortened lifespan.

Clk (**cl**o**ck** abnormal): Clk mutants display slowed temporal processes, such as a delay in larval development, and an extended lifespan compared to wildtype worms. An assay for this phenotype is described in Section 8.5.2.

Daf (**da**uer larva **f**ormation abnormal): Worms displaying Daf phenotypes improperly regulate entry into the dauer larva stage: Daf-c phenotypes are ones in which dauers are constitutively formed (i.e. in the presence of food); Daf-d phenotypes are ones in which dauer larvae do not form following appropriate cues (i.e. starvation). Assays for these phenotypes are described in Section 8.5.6.

Gro (**gro**wth rate abnormal): Worms with a Gro phenotype show a very reduced growth rate.

Hid (**h**igh temperature-**i**nduced **d**auer formation): Constitutive dauer formation at higher temperatures (i.e. 27 °C).

Liv (long-**li**ved and **v**iable after thermal stress): Worms display a lengthened lifespan following heat-shock stress.

Old (**o**verexpression **l**ongevity **d**eterminant): Worms displaying the Old phenotype show increased resistance to heat and UV stress and an increased lifespan.

Sle (**sl**ow **e**mbryonic development): Sle embryos develop at a delayed rate compared to wildtype embryos.

Morphological phenotypes

Amo (**a**bnormal **mo**rphogenesis): Worms displaying an Amo phenotype show gross defects in body formation. Generally these phenotypes can be traced to defects in cells arising from the AB lineage; often they result in embryonic or larval death.

Aph (**a**nterior **ph**arynx defective): Worms lack the anterior pharynx (the first visible bulb in the head of the animal).

***Apx** (**a**nterior **p**harynx in e**x**cess): Worms display extra cells in the anterior pharynx.

Bli (**bli**stered): Adult-specific phenotype in which the cuticle of the animal is covered with fluid-filled swellings that seem to result from separation of the cortical and basal layers of the cuticle (see Figure 8.2).

Clr (**cl**ea**r**): Fluid appears to fill the pseudocoelomic cavity such that the intestine seems to float within this fluid.

Col (**col**lagen): Worms with Col phenotypes have abnormal cuticles due to defects within the collagen matrix making up this structure. Col phenotypes may be stage-specific.

Dda (**d**umpy **da**uer): Worms displaying a Dda phenotype form short dauers.

Dig (**di**splaced **g**onad): In Dig animals, the gonad is untethered to the body wall and thus frequently becomes displaced from its normal ventral mid-body position.

***Dim** (**di**sorganized **m**uscle): Wildtype bodywall muscles are not observed in Dim animals when examined under polarized light. However, Dim mutants may move using a wildtype sinusoidal pattern.

Dpy (**d**um**py**): Dpy animals are shorter and fatter than wildtype worms of the same age (see Figure 8.2).

Dyf (abnormal **dy**e-**f**illing): Sensory amphids and phasmids develop abnormally in Dyf animals, resulting in defective uptake of dyes such as FITC, and often result in abnormal chemotaxis.

Epi (abnormal **epi**thelia): Phenotype of abnormal morphology caused by defective development of epithelial tissues (epidermis, gonad, intestine).

Evl (abnormal **e**version of **vul**va): Abnormal development of the uterus in which the final vulva eversion generates an overt protrusion at the vulval slit (see Figure 8.2). Same as Pvl.

*****Exc** (**exc**retory canal abnormal): Class of phenotypes in which the structure of the excretory canal is abnormal. These include partial or total blockage of the canal by bloated cell bodies within the lumen, canals of variable length that meander along the body of the worm, or shortened canals.

Gex (**g**ut on **ex**terior): Gex mutants result in a failure of the earliest movements of the cells forming the epidermal layer during embryogenesis.

Gla (**g**erm **l**ine **a**poptosis abnormal): Worms displaying a Gla phenotype have increased levels of apoptosis of germ cells, as examined by acridine orange staining of the gonad.

*****Gld** (defective in **g**erm **l**ine **d**evelopment): Germ cells enter but fail to exit meiosis, resulting in a gonad full of abnormally shaped cells.

*****Glo** (**g**ut granule **lo**ss): Glo mutants display mislocalization of the granules normally found within the gut of *C. elegans*, indicating a likely defect in lysosome-related organelle formation.

Glp (**g**erm **l**ine **p**roliferation abnormal): Worms with a Glp phenotype make very few, if any, germ cells. Glp animals may, but do not always, display additional somatic phenotypes.

*****Gob** (**g**ut **ob**structed or defective): Malformation of the intestine that is severe enough to prevent entry of ingested bacteria.

Gon (**gon**ad development abnormal): Early divisions of gonadal precursors and their descendants are delayed, incomplete, or lacking, resulting in a reduced or absent gonad.

*****Gum** (**gu**t **m**orphology abnormal): Gum mutants show the accumulation of large vesicles within the intestines.

Hmp (**h**ump**back**): Failure in embryonic elongation, resulting in the formation of bumps along the dorsal length of animals.

Lon (**lon**g): Animals are about 50 percent longer than wildtype; this phenotype may be visible at all stages of development or only in adult animals.

Mor (**mor**phological: rounded nose): The heads of Mor animals are less pointed than wildtype, and appear particularly round and blunted at **25×** magnification.

Mua (**mu**scle cell **a**ttachment abnormal): Worms experience progressive paralysis due to defects in muscle-to-cuticle attachment.

Muv (**mu**ltivulva): Divisions in extra ventral hypodermal cells that generate up to five pseudovulvae that appear as protrusions along the ventral side of the animal.

Nob (**k**nob-like posterior): Hatching L1 larvae with a Nob phenotype have normal head and pharynx structures but grossly misformed posteriors. Some mutations can lead to death of the larvae.

*Nog (**no** **g**ermline): In worms with a Nog phenotype, the somatic gonad develops correctly but no germ cells are formed within the gonad.

Pat (**p**aralyzed **a**rrest at **t**wo-fold stage): Pat larvae stop elongating during the two-fold stage after the embryonic body wall muscles become paralyzed. Embryos hatch as misshapen larvae.

*Pha (defective **pha**rynx development): Class of phenotypes in which the pharynx is abnormal. In some cases, the pharynx is completely absent; in others, it is misshapen.

Plg (copulatory **plg** formation): Plg males lay a gelatinous blob over the vulva of hermaphrodites following mating.

*Pro (**pro**ximal proliferation in germline): Disruption in germline progression in which the gonad contains an ectopic region of proliferating cells near the vulval region of the animal.

Pvl (**p**rotruding **v**u**l**va): Vulval tissue extends from the ventral side of the animal due to either a defect in cellular adhesion or defects in cell division in the vulval lineage cells. Same as Evl.

Rol (**rol**ler): Animals are helically twisted and unable to move in sinusoidal wildtype pattern; instead they roll on their side. This phenotype is associated with defects in collagen.

*Shv (**sh**i**v**a (four-armed) gonad): Hermaphroditic gonad develops abnormally with multiple arms, often with excess distal tip cells.

Sqt (**squa**t): Morphological defect in which animals are slightly dumpy or may be rollers. Both types of mutants may arise from a single parent.

*Sqv (**sq**uashed **v**ulva): Animals with a Sqv phenotype display a partial collapse of the vulval invagination and elongation of the central vulval cells.

Stu (**st**erile and **u**ncoordinated): Animals are both uncoordinated and unable to produce offspring.

Vab (**v**ariable **ab**normal morphology): A phenotypic class in which animals display abnormal body shape due to defects in morphogenesis during development.

*Vul (**vul**valess): Loss of cells that generate modified epidermal structure used for laying eggs.

Embryonic phenotypes

*Cdl (**c**ell **d**eath **l**ethal): Worms die during embryonic development and display an accumulation of cell corpses within the embryo.

*Ced (**ce**ll **d**eath abnormality): Programmed cell death is abnormal in Ced animals, such that dying cells arrest in a highly refractile stage and killer cells fail to engulf their target cells properly.

*Cow (**co**ntractile **w**aves in embryo): Embryos display excessive cortical contractions throughout the one-celled embryo as opposed to waves restricted to only the anterior cortex.

*Cta (**c**y**t**oplasmic **a**ppearance): Embryonic lethal phenotypes in which embryos have large areas containing no yolk granules.

*Eld (**el**ongation **d**efective): Lethal phenotype in which developing embryos fail to elongate beyond the comma stage.

Emb (**emb**ryogenesis abnormal): Class of phenotypes in which embryonic development is abnormal and often arrests. These may include defects in early cell division events, leading to arrest or defects in cell lineages and then to lethality. However, Emb defects are not always lethal.

*Lit (**l**oss of **i**n**t**estine): Lit animals lack intestinal cells and instead carry extra pharyngeal tissue. This phenotype is caused by developmental defects during embryogenesis.

Ooc (**oo**cyte formation abnormal): Lethal phenotype in which animals lay eggs that fail to hatch. In some Ooc animals, abnormal oocytes are visible within the gonad.

*Pod (**p**olarity and **o**smotic sensitivity **d**efective): Embryos display defective anterior/posterior polarity and are osmosensitive due to apparent defects in the eggshell.

Zyg (**zyg**ote defective): A large class of phenotypes that always result in embyronic lethality.

Mutant-specific phenotypes

Chb (**ch**e-2 small **b**ody size suppressor): In *C. elegans*, body size is at least partially regulated by the nervous system. One class of neural mutants, the Che genes (see below), often display a smaller body size compared to wildtype. The Chb genes suppress this smaller body size (worms are wildtype in size).

Elm (**e**nhancer of **l**in-12 (gf) **m**ultivulva phenotype): Elm phenotypes result from cell division defects that yield more than the normal six pseudovulvae found in *lin-12* animals.

Scd (**s**uppressor of **c**onstitutive **d**auer formation): Scd mutants prevent dauer formation in mutant strains that normally form dauers constitutively.

Sdf (**s**ynthetic **d**auer **f**ormation): Sdf phenotypes are caused by mutations in a second gene that contributes to changes in the normal dauer formation pattern of a Daf mutant strain.

Smu (**s**uppressor of **m**ec and **u**nc defects): Phenotype in which a second mutation suppresses Mec or Unc phenotypes.

Spy (**s**uppressor of **p**olyray): Phenotypic class in which a second defect suppresses the ectopic ray formation of the Pry phenotype; it may generate a protruding vulva in hermaphrodites.

Srl (**s**uppressor of **r**oller **l**ethality): Mutations that overcome the lethality of some roller mutations but not the Rol phenotype.

Sum (**su**ppressor of **m**ultivulva phenotype): Class of phenotypes in which a second mutation overcomes a multivulva mutation.

Sex-specific phenotypes

Cod (**co**pulation **d**efective): Cod mutants are morphologically normal males that are unable to mate successfully.

*Fem (**fem**inization of XX and XO animals): Feminized worms do not produce sperm but do produce normal, fertilizable oocytes.

*Fog (feminization of germline): Mutations that prevent the formation of sperm and instead convert these germ cells to oocytes.

Her (hermaphroditization of XO animals): Males are transformed into fertile hermaphrodites.

Him (high incidence of males): Increased incidence of chromosomal non-disjunction, leading to the formation of a high number of males in a population.

*Lep (leptoderan male tail): The tip of the male tail is pointy and protrudes beyond the posterior edge of the tail fan.

Lov (location of vulva defective): Defect in male animals in which they are unable to find the hermaphroditic vulva during mating.

Mab (male abnormal): Class of defects in which the male tail is grossly abnormal in morphology.

*Mog (masculinization of germline): Mog mutants fail to switch from spermatogenesis to oogenesis during gonadal development. Instead, animals make sperm for their entire life cycle.

*Pry (polyray): Males with the Pry phenotype do not form alae but instead contain ectopic rays.

*Ram (ray morphology abnormal): Males have lumpy, amorphous rays in their tails.

Smg (suppressor with morphogenetic effect on genitalia): Adult animals display abnormal genitalia: males have swollen tails and hermaphrodites have protruding vulvae.

Tra (transformer): Genotypic hermaphrodites are converted to male animals of low or no fertility.

Miscellaneous phenotypes

Let (lethal): Phenotypic class in which larvae die during development.

Osr (osmotic stress related): Osr animals are capable of retaining body water content and movement when placed in extreme osmotic stress conditions that ordinarily dehydrate animals.

Appendix V   Vector maps

Vector **pLexA** is a yeast two-hybrid DNA-binding domain vector containing the MCS illustrated in Figure AV.1. It is used for traditional cloning of a target cDNA sequence for a yeast two-hybrid screen. Vector **pLexAgtwy** is a Gateway-modified version used for Gateway cloning of a cDNA to be used as the bait in a yeast two-hybrid screen. Gateway cloning is outlined in Appendix I. This vector was modified from pLexA by placing a Gateway acceptor cassette in the Sma I site of pLexA. To facilitate subcloning of a given DNA insert into this plasmid, Gateway recombinational attachment sites should be incorporated into primers used for amplification, as outlined in the Invitrogen Gateway manual. Whereas pLexA itself can be propagated in any *E. coli* strain (using ampicillin resistance as a marker), pLexAgtwy is propagated only in *E. coli* strain DB3.1, due to the requirement to circumvent the lethality of the inherent ccdB gene until an insert is recombined into the Gateway cassette.

Vector **pACT2.2** is a yeast two-hybrid transcriptional activation domain vector containing the MCS illustrated in Figure AV.2. It is used for traditional cloning of a target cDNA sequence to test a directed yeast two-hybrid interaction (no screening involved). Vector **pACT2.2gtwy** is a Gateway-modified version used for Gateway cloning of a cDNA to be used to test a directed yeast two-hybrid interaction (no screening involved). This vector was modified from pACT2.2 by placing a Gateway acceptor cassette in the Sma I site of pACT2.2. To facilitate subcloning of a given DNA insert into this plasmid, Gateway recombinational attachment sites should be incorporated into primers used for amplification, as outlined in the Invitrogen Gateway manual. Whereas pACT2.2 itself can be propagated in any *E. coli* strain (using ampicillin resistance as a marker),

Integrated Genomics   Guy A. Caldwell, Shelli N. Williams, Kim A. Caldwell
© 2006 John Wiley & Sons, Ltd

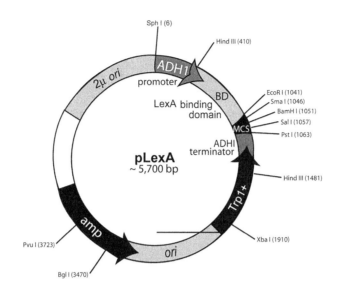

MCS: GAA   TTC   CCG   GGG   ATC   CGT   CGA   CCT   GCA   G

EcoR I   **Sma I**   **BamH I**   **Sal I**   **Pst I**

Figure AV.1   Detailed diagram of vector pLexA

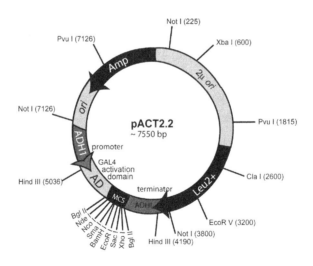

MCS:

CAT ATG GCC ATG GAG GCC CCG GGG ATC CGA ATT CGA GCT CGA GAG ATC T

Nde I   **Nco I**   **Sma I**   **BamH I**   **EcoR I**   **Sac I**   **Xho I**   **Bgl II***

*this site is not unique

Figure AV.2   Detailed diagram of vector pACT2.2

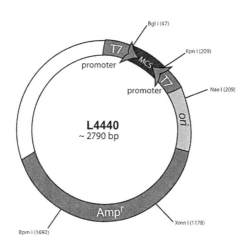

T7 promoter CCGGCAGATCTGATATCATCGATGAATTCGAGCTCCACCGCGGTGGCGGCCGCTC

TAGATCTAGAACTAGTGGATCCACCGGTTCCATGGCTAGCCACGTGACGCGTGGATCCCCCGGGCTGCA

GGAATTCGATATCAAGCTTATCGATACCGTCGACCTCGAGGGGGGGCCCGGTACCCAATT **T7 promoter**

Figure AV.3 Detailed diagram of vector L4440

pACT2.2gtwy is propagated only in *E. coli* strain DB3.1, due to the requirement to circumvent the lethality of the inherent ccdB gene until an insert is recombined into the Gateway cassette.

Vector **L4440** is an RNAi feeding vector containing the MCS illustrated in Figure AV.3. It is used for traditional cloning of an RNAi target sequence for RNAi feeding experiments. Vector **L4440gtwy** is a Gateway-modified version used for Gateway cloning of an RNAi target sequence as outlined in Appendix I. A Gateway recombinational cassette was placed into the EcoRV site of L4440 to generate this modified plasmid. To facilitate subcloning of a given DNA insert into this plasmid, Gateway recombinational attachment sites should be incorporated into primers used for amplification, as outlined in the Invitrogen Gateway manual. Whereas L4440 itself can be propagated in any *E. coli* strain (using ampicillin resistance as a marker), L4440gtwy is propagated only in *E. coli* strain DB3.1, due to the requirement to circumvent the lethality of the inherent ccdB gene until an insert is cloned into the MCS.

Subject index

Note: page numbers in *italics* refer to figures and tables

Integrated Genomics   Guy A. Caldwell, Shelli N. Williams, Kim A. Caldwell
© 2006 John Wiley & Sons, Ltd

Index compiled by Neil Manley